First published 1978 and this revised edition 1980, 1981 by Writers and Readers Publishing Cooperative Limited, 9-19 Rupert Street, London W1, England

Text copyright ©1978, 1980 Stephen Croall *Illustrations copyright* ©1978, 1980 Kaianders Sempler

Series Editor Richard Appignanesi
Additional Design and Layout Safu-Maria Gilbert
Cover Design Louise Fili
Typesetting in 14-16 Helvetica by Fi Litho and Range Left, London

PRINTED IN CANADA

ISBN 0 906495 62 8 (Cased)
ISBN 0 906495 63 6 (Paper)

Nuclear Power

FOR BEGINNERS

Stephen Croall and Kaianders Sempler

Writers and Readers

6

HERE COMES
CHAPTER 1.

In which the meaning of energy remains a mystery and history is reduced to a quick flit from wood-burning to Crazy Joe's Massacree...

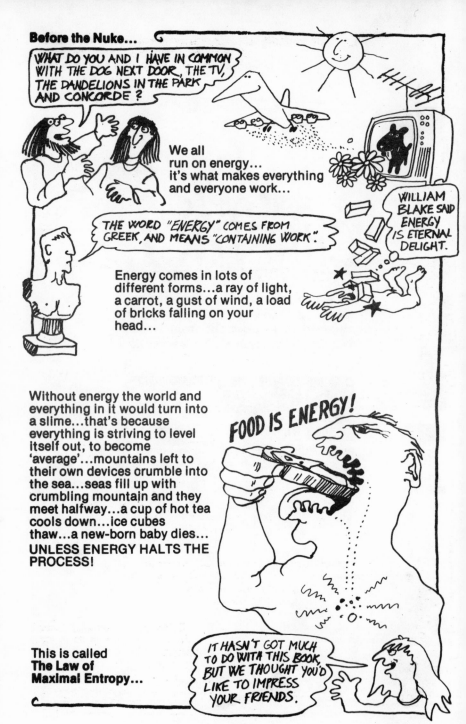

WHAT DO YOU AND I HAVE IN COMMON WITH THE DOG NEXT DOOR, THE TV, THE DANDELIONS IN THE PARK AND CONCORDE?

We all
run on energy...
it's what makes everything
and everyone work...

THE WORD "ENERGY" COMES FROM GREEK, AND MEANS "CONTAINING WORK".

WILLIAM BLAKE SAID ENERGY IS ETERNAL DELIGHT.

Energy comes in lots of different forms...a ray of light, a carrot, a gust of wind, a load of bricks falling on your head...

Without energy the world and everything in it would turn into a slime...that's because everything is striving to level itself out, to become 'average'...mountains left to their own devices crumble into the sea...seas fill up with crumbling mountain and they meet halfway...a cup of hot tea cools down...ice cubes thaw...a new-born baby dies... **UNLESS ENERGY HALTS THE PROCESS!**

FOOD IS ENERGY!

This is called **The Law of Maximal Entropy...**

IT HASN'T GOT MUCH TO DO WITH THIS BOOK, BUT WE THOUGHT YOU'D LIKE TO IMPRESS YOUR FRIENDS.

Until recently all our energy came from the sun...
directly or indirectly...it pours down
far more energy than the human race can
ever hope to use...

SO WHAT'S ALL THE TALK ABOUT AN ENERGY CRISIS?

But to start from the beginning (whenever that was)...the first men and women only had food as a source of energy. Tired of jumping up and down to keep warm they lit a fire...so wood was the first fuel...

WOW!

Everyone helped themselves generously to what looked like an endless supply of energy...and by the Christian Era the forests of northern Africa and the Middle East had all but vanished — gone to fuel, building timber and farmland...

TODAY WE KNOW THESE REGIONS AS DESERT.

People shouldered their bags, moved on to Western Europe and spent the Middle Ages cutting down all the trees there instead...

Strangely, they didn't start clearing the North American continent until modern times ...and there's still some virgin forests in Siberia and the amazons...

UNITED LOGS

YEAH, WE'RE GETTING ROUND TO THEM...

9

After wood came wind and water...well, they'd been there for some time but people hadn't realised they could be used to make things move...when that idea hit home, there was a scramble to knock up windmills, wind pumps, watermills, sailing ships and so on...

Sounds fine, but...the wind also blew the ships of the 'civilised countries' far afield and enabled them to plunder, enslave and colonise 'uncivilised countries'

The British were so good at this particular sport that they ran out of timber for shipbuilding...luckily they had lots of coal to use as fuel...so they began mining it on a big scale...and sparked off **THE INDUSTRIAL REVOLUTION**...

OTHERWISE KNOWN AS THE BIRTH OF CAPITALISM...

COAL and the steam-engine launched the machine age in the 18th century…by the middle of the 19th OIL & GAS were also being brought out of the ground…

Thus the Industrial Revolution heralded a new environmental era — the speedy exploitation of an energy supply that could not be renewed…FOSSIL FUELS…

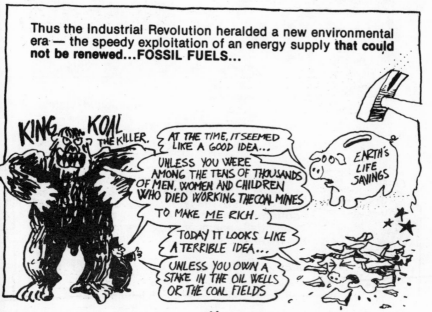

The sun spent 600 million years fossilising marine life into oil and trees into coal...and at the rate we're going today they'll be used up within the next few hundred years..

NO MORE OIL...

NO MORE COAL...

GOOD-BYE GAS...

1700   1800   1900   2000

NON-RENEWABLE SOURCES OF ENERGY

like coal, oil and natural gas — and uranium, the raw material for nuclear power — have certain things in common that appeal to ruling minorities...

THEY'RE CONCENTRATED IN TERRITORY WHICH CAN BE "OWNED"...

LOTS OF MONEY IS NEEDED TO BRING THEM UP AND PROCESS THEM

A HIGH DEGREE OF TECHNICAL KNOW-HOW IS REQUIRED

THAT'S GRATITUDE FOR YOU!

£ 1.000,000

RENEWABLE SOURCES OF ENERGY

— sun, wind, water & plants — are something else...

IT'S NOT EASY TO CORNER THE MARKET IN SUNSHINE...

They're almost impossible to exhaust...they're available in many parts of many countries...and they're potentially easier and cheaper to harness...

Far too democratic for some...

..OR CENTRALIZE WIND...

SAM

IVAN

12

TO CUT A LONG STORY SHORT...

electricity arrived at the end of the 19th century and with generous helpings of oil and coal the industrialists steamed into the 20th hell-bent on bigger factories, better machines, deeper mines...and, of course, larger profits...

This was in the 'civilised' parts of Europe and America...

...the rest of the world paid the bill...

JUST BORROWING THEM...

NATURAL ASSETS AND RESOURCES

At the end of World War I the West lost a market, Russia going commie in 1917...

To divert attention from this breakthrough and keep the wheels of capitalism turning a US industrialist, H. Ford, had a bright idea:

EVERYONE SHOULD HAVE A MOTOR-CAR!

YOU DIRTY RED!

At the start of the 1930s...

THE BUBBLE FINALLY BURST

Ford and his fat friends in the US, Britain, France and Germany had been so busy lining their own pockets that no money was left over for anyone to buy their goods...

WANNA BUY? WANNA BUY?

SURE. DO YOU TAKE COLOURED BEADS?

They closed down their factories and banks and ran for the hills...leaving some 40 million people jobless, penniless and bitter...and laying the ground for fascism and World War II...

Having already kicked out the capitalists, the Soviet Union escaped this Great Depression...but it was having troubles of its own, with Crazy Joe Stalin wiping out millions of peasants who objected to his way of doing things...

HEY... I THOUGHT THIS BOOK WAS ABOUT NUCLEAR POWER!

BLAM!

OK, how's this...

MEANWHILE, BACK AT THE LAB...

THIS IS CHAPTER 2.

In which the atom is split by mistake, some 200,000 Japanese are killed on purpose, Sam & Ivan play in the sandpit and the Indians call everyone's bluff...

Scientist are funny people...

Give them an **atom**, tell them it's the smallest thing around and what do they do? Split it in two, of course. Just to rub it in, they'll split the very heart of the atom — the **nucleus!**

The scientists spent the first three decades of the 20th century probing the nature of the atom...

NIELS BOHR OF DENMARK **ERNEST RUTHERFORD** OF BRITAIN

Unlocking the atom, though, proved a tough nut until a Briton called **James Chadwick** discovered a tiny particle in 1932 — the **neutron**...

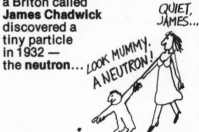

By bombarding atoms with neutrons scientists found they could turn one chemical into another...and this became a favourite pastime in the labs of the 1930s...

Among the most skilful neutron-tossers was the Italian Enrico Fermi...

**One day they tried bombarding the nucleus of a uranium atom...
and it disappeared! Something was up...**

OTTO HAHN.

In 1939 an Austrian physicist, Lise Meitner, rescued the scientific community from going crazy by suggesting the uranium atom had *split in two*...a process she called **FISSION**

Ms Meitner's decisive role in this historic achievement was rewarded with a Nobel Prize...for her ex-partner Otto Hahn!

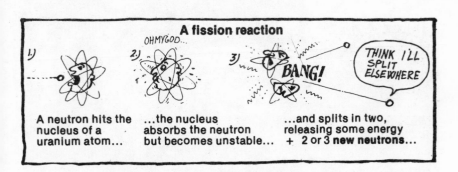

A fission reaction

A neutron hits the nucleus of a uranium atom...

...the nucleus absorbs the neutron but becomes unstable...

...and splits in two, releasing some energy + 2 or 3 **new neutrons**...

The release of fresh neutrons opens up the prospect of a CHAIN REACTION...

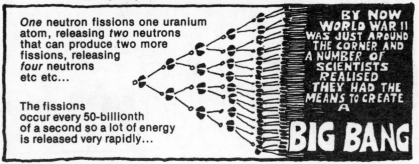

One neutron fissions one uranium atom, releasing *two* neutrons that can produce two more fissions, releasing *four* neutrons etc etc...

The fissions occur every 50-billionth of a second so a lot of energy is released very rapidly...

BY NOW WORLD WAR II WAS JUST AROUND THE CORNER AND A NUMBER OF SCIENTISTS REALISED THEY HAD THE MEANS TO CREATE A **BIG BANG**

17

US President Roosevelt was told that if the Western Allies didn't stick that chain reaction inside a weapon quickly the mad A. Hitler would get the bomb first...so in 1941 he gave the go-ahead for

The Manhattan Project

The cream of US, British, Canadian and French scientists got together to make a nuclear fission bomb...

Fermi, the Italian whizz kid, played a key role.
Some have called him "the greatest physicist of the century". Others remember his stock response when colleagues muttered about the A-bomb posing a threat to humanity:

DON'T BOTHER ME WITH YOUR CONSCIENTIOUS SCRUPLES. AFTER ALL, THE THING'S SUPERB PHYSICS!

BOUM
BANG!

$E = mc^2$
$^{235}_{92}U$ $^{239}_{94}Pu$
$\alpha + \beta + \gamma + n$

SIEG HEIL

BANG!

Fermi & Co reached their first goal in December 1942. A clumsy uranium reactor built in a converted squash court under a Chicago sports stadium worked at the first try...

The energy it produced would not have lit a pocket torch...but it marked Planet Earth's entry into **THE ATOMIC AGE**...

AWFULLY SECRET

I STILL DON'T UNDERSTAND WHY FRANKIE WANTED TO BOMB MANHATTAN...

F.D.R

By now US scientists had turned up another fissionable element that was even better than uranium for making bombs... PLUTONIUM...one of the deadliest substances ever created by Man.

ALL NUCLEAR REACTORS BURN URANIUM AND PRODUCE PLUTONIUM!

Over the next couple of years the bomb materials — plutonium and *enriched* uranium — slowly mounted up ...the project was so hush-hush that Harry Truman didn't even know about it when he took over the presidency on Roosevelt's death in 1945...

A New York journalist who nosed up evidence that nuclear energy could make an almighty big blast discovered later that his articles had been classified top secret — after publication.

DAILY SNOOPER
BIG BANG CAN WIN WAR!
BANG URANIUM

By the time they were ready to test, in July 1945, some of the scientists were having serious doubts about the morality of a nuclear fission bomb. But they were outnumbered. The test, at **Alamagordo** in the New Mexico desert was a shattering success and terrified some of the bomb's creators. Asked to describe what he'd seen, one physicist replied: "I can't tell you, but . . .

DON'T EXPECT TO DIE A NATURAL DEATH!

CLAP CLAP

The Manhattan Mob lost no time in demonstrating to the world how greatly they'd improved the human race's capacity to destroy itself...on August 6, 1945, a uranium bomb nicknamed *Little Boy* was dropped on the Japanese city of Hiroshima. It worked and the war was as good as over... But the boys back home were anxious to try out their plutonium bomb as well... So three days later *Fat Man* was dropped on the city of Nagasaki... That one also worked.

The two raids killed some 200,000 people immediately. Another 100,000 died from the after-effects of radiation over the ensuing decades. Today, victims of **Little Boy** and **Fat Man** are still dying in Japan at the rate of 2,000 a year... .

**World War II came to an abrupt end...
The development of nuclear weapons did not...**
The US Navy was sulking because
the US Army had gotten all the
glory for outstanding services
to mankind at Hiroshima and
Nagasaki...

The marines
needed
to
restore
their
manhood...

So in the summer of 1946
they arranged for 42,000
spectators to witness two bangs
at a Pacific atoll
called Bikini...

THE BIKINI ATOLL.

Absent from this display of
heroics were the local islanders...
they'd been secretly removed a few months earlier...
and could not return until 22 years later...by which time the
vegetation had become unproductive and the menu was
radioactive coconut crab...

PEACE
FOR
PRODUCTION
PROGRESS
PROFIT.

MEANWHILE,
Back in the States, the
politicians and businessmen
were getting excited. Uncle
Sam was the only one **in the
whole wide world** with an
A-bomb! Time to spread some
Pax Americana around the
globe by gentle persuasion...
And not a moment too soon,
what with the weirdo commies
fanning out in Eastern
Europe and the **Yellow Peril
going red in China...**

JUST TO BE ON THE SAFE SIDE
Sam passed the **McMahon Act.**
This neat little law meant
that anyone who thought
the Manhattan project
had been a joint effort
was out of his mind.
The thing had been built
in Sam's back yard
so it was only fair that
all nuclear information
be labelled:

The **British**, of course, were absolutely furious...They still hadn't got over the way the bloody Yanks had taken all the bloody credit for winning the bloody war...**Now Sam was showing them the nuclear door!** They stormed home in a huff to build their own bangs...

The **Canadians** were
so upset they
decided not to make
any bombs at all...

**So the stage was set
for the United States of the World...**
or so it might have seemed to some...
Then **Ivan's Gang** went and spoiled
everything by
testing a fission
bomb of their own in
September 1949.
It was six times
bigger than
Little Boy...

THE RUSSIANS ARE COMING!

Now it was common knowledge that no goddam
Russky could get the better of Sam except by cheating.

Scapegoats were soon found —
a couple called Rosenberg
who were accused of slipping
Ivan **the secret formula...**
In the thrill of a trial and
execution most people
overlooked
the fact that the
'secret formula' had long
been in general circulation in the scientific community...and that
Ivan had been fully engaged in nuclear research since 1940...

The Soviet mushroom launched
the Arms Race...also known as
The Sandpit Syndrome...

MINE IZ BIGGER! HE HE...

HO HO, MINE CAN GO FURTHER! HAHAHA!

BANG!

WHO SLIPPED THEM THE SECRET?

In October 1952 Britain's military scientists stopped sulking long
enough to blow the fish out of the water outside Australia...

London newspapers started tossing around phrases like **The Big Three...** but no-one was really fooled... fission bombs were kid's stuff by now and the two superpowers were working on superbombs combining fission with **fusion...**

Fusion is slamming together **hydrogen nuclei***to form **helium** and create a **thermo-nuclear chain reaction...**

Whereas the fission bomb could destroy cities the H-bomb could destory whole regions.

Sam gave it a ground run in November 1952, wiping a Pacific island off the map with a blast 500 times bigger than Hiroshima...

* One nucleus, two nuclei. Latin form. Ask any of the old Romans.

Ivan dropped an even bigger one in August 1953...which Sam topped in March 1954... but made himself a bit unpopular by raining the radioactive fallout on a couple of hundred islanders 150 kilometres away...and on the Japanese fishing vessel *Lucky Dragon*...

One of the crewmen died.

Thus the first victim of the **H-bomb** was Japanese, as were the first victims of the **A-bombs**... So it's not surprising that US repairmen often have to be flown in when nuclear power stations go wrong in Japan today...the natives are a bit sensitive about radiation levels...

MEANWHILE,

Sam was at war again — disguised as
a United Nations soldier — trying to
win Korea for the Free World...
Leading the charge was World War II
hero **General MacArthur**, who was so
keen on liberating the place that
he asked for atomic bombs...

President Truman didn't think
the idea sounded too bad...but he
was talked out of it by his mates
at the UN...and Mac lost his job...

Anxious as ever to play with the
big boys, Britain let off an H-bomb
in 1957...and over the next decade
France and China tested atomic and
hydrogen bombs to join the
Fission & Fusion Freaks...

3-2 TO THE CAPITALISTS

Why stop at the H-bomb?
What about a weapon capable of
blowing the world to bits?
Scientists have considered
this tempting project...
on the principle that the
pursuit of truth knows no
boundaries...but
ruefully concluded that
the **Doomsday Bomb** is a
technical impossibility...

27

By way of consolation Sam's military boffins gave us:

THE NEUTRON BOMB

First tested in 1977, this delightful bit of hardware has one major advantage — **it does little damage to buildings...** It just spews out neutrons (*remember the neutrons?*) and kills all life within range...any survivors are thus spared the bother of building up a new production apparatus...

But we're getting ahead of the story...

The Arms Race accelerated into the 1970s with Sam and Ivan battling for the lead, the Chinese a good third and the French and British gamely hanging on...

Suddenly out of nowhere came the Indians!

It was all terribly embarassing.

For years we'd been assured by
       the scientists
       the military
       the industrialists
       the politicians

PEACEFUL NUCLEAR TEST.

AGRESSIVE NUCLEAR TEST.

Had someone blown their cover 10 years earlier things might have turned out differently...

But by the end of Indian Bomb Year more than 150 of those harmless little nuclear power plants (*nukes*) were sprouting up around the globe...

The 'exclusive' Nuclear Arms Club now looks like getting a dozen new members during the 1980s...

30

NOW ON TO
CHAPTER 3.

In which «peaceful» nukes spread around the globe, the oil giants become energy giants, the Human Factor gets into his stride and accidents tend to happen...

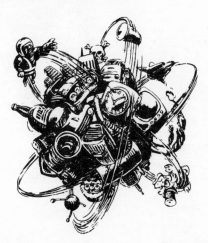

37658

Originally the heat generated in the fission process was considered a useless by-product...

But by the late 1940s a lot of people realised reactors were more than just a way of producing plutonium...

Some people got carried away...

To the European industrialists nuclear power sounded like a godsend. They were running short of cheap fuel just when a Second Industrial Revolution was in the offing. The fact that nukes generate electricity and that electricity has only limited uses wasn't mentioned much...

The military had set the wheels in motion, laying on investment, research and development of nuclear power ...the capitalists just had to climb aboard...

32

Over at Sam's place private industry wasn't into nukes...there was plenty of cheap oil and the prospect of lots of coal and natural gas...

It quickly changed its mind when the **federal government** threatened to go into the business...

THOSE WHITE-HOUSE COMMIES HAVE GOT TO BE STOPPED !

Canada was the first country to concentrate on civil nuclear programmes...and, in 1952, the first to play host to a major accident. A research reactor at Chalk River exploded but no-one was killed so no-one got excited...

WE'RE PROUD TO ANNOUNCE THE FIRST *INDOOR* NUCLEAR EXPLOSION.

Sam & Ivan boasted sizable reactors by 1954...but Britain (Calder Hall) and **France** (Marcoule) claimed the first **commercial** nukes in 1956, generating electricity to the public...

> THEY ALSO GENERATE PLUTONIUM FOR US...

> NAM NAM...

> I SEEM TO HAVE STARTED SOMETHING...

> ...THAT I CAN'T STOP.

> IAEA

In the same year the **International Atomic Energy Agency [IAEA]** was set up in Vienna. It aimed to **promote the spread of nuclear power** and **prevent the spread of nuclear weapons materials.** Like teaching a child to walk then ordering it not to...

Sam took a short cut...in 1957 he called home the nuclear-powered submarine **Nautilus,** stole its reactor and set up the first US nuclear power station at Philadelphia...

> NAUTILUS SHIPPINGPORT

Ivan built *his* first civil nuke at Troitsk in Siberia...a breakthrough for Leninism??

> COMMUNISM IS SOVIET POWER PLUS ELECTRIFICATION OF THE WHOLE COUNTRY.

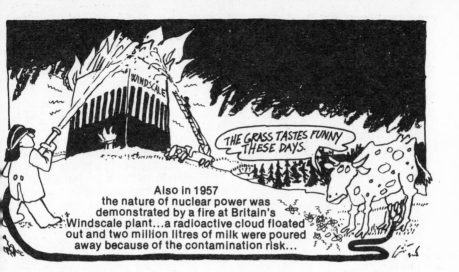

Also in 1957
the nature of nuclear power was
demonstrated by a fire at Britain's
Windscale plant...a radioactive cloud floated
out and two million litres of milk were poured
away because of the contamination risk...

By this time, several
countries *that did not
have nuclear arms* were
charging head-first into
the nuclear age...West
Germany...Canada...Italy
...Japan...Sweden...

All the nuclear states had one thing in common — they'd taken the
plunge without giving the general public a chance to express an opinion
about nuclear power. The scientists had consulted the military, the
military had consulted the politicians, who consulted the businessmen,
who had a word with the scientists, and round and round they went...

BUT FOR SOME REASON NONE OF THEM
WAS INTERESTED IN CHECKING IT OUT WITH
THE PEOPLE.

THE WHAT ??

In fact, not until 30 years after nukes made their appearance were the people of any country asked in any way whatsoever if they wanted this extremely dangerous technology...

IT'S FAR TOO COMPLICATED FOR ORDINARY MORTALS.

Today, so much state and private capital has been poured into nuclear power expansion that a critical public is faced with a massive wall of inter-locking economic interests and misleading propaganda...

BACK TO HISTORY. The 1960s opened with a bang — at the Idaho Falls SL-1 nuclear plant in the US. Three repairmen were killed...

Their bodies were so contaminated by radiation that they couldn't be buried for 20 days...and then only in lead-lined coffins sealed in lead-lined tombs...

ROAD CLOSE

The accident was blamed on human error...as were the accidents at Chalk River and Windscale...**and most of the mishaps that were to plague every major reactor design before the end of the decade!**

IN FACT, HUMAN BEINGS AND NUCLEAR POWER SEEMS A LETHAL COMBINATION.

The worst reactor accident to date may or may not have been the one in 1963 involving Sam's nuclear submarine **Thresher.** It disappeared on a deep dive with 129 people aboard. No-one knows for sure whether it fell victim to a reactor breakdown...

Three years later came the confrontation everyone had been dreading...**the Human Factor met the Breeder Reactor!**

They crossed paths at the Enrico Fermi experimental plant outside Detroit. Someone forgot to fasten a bit of metal in the reactor core, it caused part of the fuel to melt and orders went out to prepare for the evacuation of the city...

A **disaster was averted**...but only just. Had it occurred, claiming hundreds or thousands of lives, the development of nuclear power would probably have halted...or at least fast breeder development...

Instead, the second half of the 1960s brought an upswing in sales. Nuclear electricity was now looking profitable so Sam's heavies decided it was time to muscle in...

The giant US firms had got off to an early start so they were well placed to seize control of the western reactor market...and they still dominate today...

ANATOMY OF AN OIL CRISIS...

In 1973 the Arab oil-producing countries jacked up basic prices by 400%...and the 'black gold' that had fuelled the rapid industrial build-up of the West suddenly became 50% more expensive to the consumer...

SO THE OIL COMPANIES WENT BUST?

ON THE CONTRARY... IT WAS JUST WHAT THEY'D BEEN WAITING FOR!

They knew they were emptying the oilfields too fast...there were nationalisation rumours all over the Middle East...and that they'd have to branch out to stay in business...but that required immense investment...and the low price of oil had meant it was unprofitable to move into coal, gas or nukes...

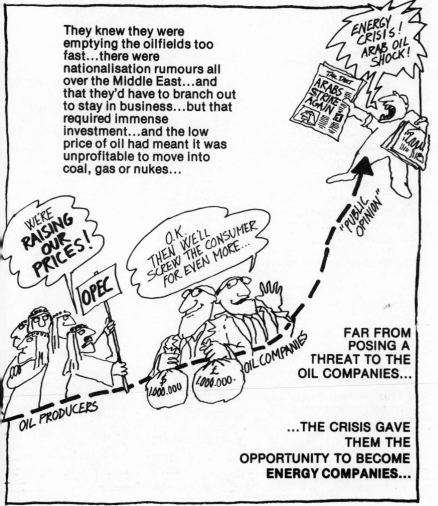

ENERGY CRISIS! ARAB OIL SHOCK!

The Diet
ARABS STRIKE AGAIN

WE'RE RAISING OUR PRICES!

OPEC

O.K. THEN WE'LL SCREW THE CONSUMER FOR EVEN MORE...

"PUBLIC OPINION"

$1,000,000   £1,000,000

OIL COMPANIES

OIL PRODUCERS

FAR FROM POSING A THREAT TO THE OIL COMPANIES...

...THE CRISIS GAVE THEM THE OPPORTUNITY TO BECOME ENERGY COMPANIES...

NUCLEAR POWER **LOOKED** THE BEST BET...

§ It promised to yield greater profits than coal

§ There was no union militancy in the nuclear industry ...while coalminers showed growing reluctance to die for the bosses

§ Uranium reserves were in politically 'safe' places like the US, Canada, Australia and...er...southern Africa

Although Sam had a firm grip on the reactor-building business by the 1970s he was not without rivals...**The British** for example had come up with a nuke they called the AGR. It was such a mess that it's no longer mentioned in polite company...

MUMMY, WHY DON'T THEY FINISH THE DUNGENESS AGR?

SSSSHH...

The Swiss also had a crack at making one...and wisely stuck it underground, at Lucens. After two years the core blew up and the reactor had to be scrapped...

THE WORLDS FIRST OIL-FIRED NUCLEAR STATION

Sweden's first attempt was even more embarrassing...it couldn't be made to work at all! The plant, at Marviken, was refitted to run on oil...

CANDU

WOW!

By contrast, *Canada's* CANDU reactor proved a money-spinner. Exactly why became clear when India startled everyone with its bomb test in 1974 — the CANDU is a very efficient producer of plutonium...

But the nukes that outsold them all were Sam's **LIGHT WATER REACTORS [LWRs]**...Governments and power companies were in such a hurry to buy that they didn't worry too much about minor details like rotten safety records...

THE HUMAN FACTOR

The 1973 'oil crisis' had opened a lot of doors to the nuclear salesmen. By the mid-1970s there were 19 countries in the nuclear fold...among them two from the Third World...

THE HUMAN FACTOR

meanwhile, was having a gay old time...he fitted one reactor back-to-front, fitted pumps upside down in another, forgot a welding rig inside a third, lost track of uranium shipments and plutonium stocks, and shut down nukes left, right and centre...

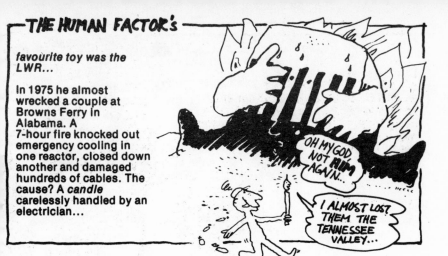

THE HUMAN FACTOR's

favourite toy was the LWR...

In 1975 he almost wrecked a couple at Browns Ferry in Alabama. A 7-hour fire knocked out emergency cooling in one reactor, closed down another and damaged hundreds of cables. The cause? A *candle* carelessly handled by an electrician...

OH MY GOD, NOT HIM AGAIN...

I ALMOST LOST THEM THE TENNESSEE VALLEY...

Despite the antics of HF the nuclear power industry predicted a dramatic worldwide expansion...with nukes in 40 or 50 nations by the end of the century meeting 50% of global electricity 'needs'...

...and the Fast Breeder just waiting to solve all our energy problems in the next century...

YUM, YUM GIMME! GIMME!

BUT SOMETHING WENT SOUR... In the second half of the 1970s a lot of people woke up to the dangers of nuclear power...and threw the whole future of the industry into doubt...

CLOSED FOR NON-PROFIT REASONS

$1,000,000

THE INDIANS HAD DEMONSTRATED THAT THERE WAS NO SUCH THING AS A **PEACEFUL ATOM**... BUT THAT DIDN'T WORRY THE INDUSTRY.

THE CLAIM THAT NUKES WERE RELIABLE AND SAFE RANG INCREASINGLY HOLLOW WITH EVERY BREAKDOWN AND DISASTER... THAT **DID** WORRY THE INDUSTRY. A BIT...

BUT NUKES BEGAN TO LOOK LIKE A **BAD ECONOMIC INVESTMENT**, AND THAT **REALLY** UPSET THE INDUSTRY!

WHEN YOURE TRYING TO MONOPOLISE THE WORLD ENERGY SUPPLIES, HIGH RISKS AND LOW PROFITS ARE THE LAST THINGS YOU NEED!

In the US where half the world's nuclear trade goes on reactor salesmen suddenly found doors being slammed in their faces...and orders being postponed or cancelled. It was all too much for the major oil companies...

WHY DO YOU REFUSE TO PUBLISH SEPARATE ACCOUNTS FOR YOUR REACTOR BUSINESS?

WELL... ER... I'M JUST NATURALLY MODEST...

Shell, Gulf & Co started shedding their reactor interests...only Westinghouse of the big corporations claimed to be making a profit in the nuclear field...

44

PROF IT

In Japan and Western Europe too, ambitious nuclear programmes were cut back severely. Almost everywhere nukes ran into political, economic or technical trouble. They became so unpopular that some governments felt obliged to hold referendums...

IN AUSTRIA WE STOPPED THE NUKE

IN SWITZERLAND WE JUST FAILED

SPECIAL BARGAIN! BUY BRITISH!

As for Britain, home of the first 'commercial' nukes, only massive state backing saved the domestic reactor industry from total collapse in the 70s...

AT MY PLACE, WE HAVE NO SUCH PROBLEMS, SO IT'S FULL STEAM AHEAD.

Suddenly, something happened that brought nuclear power to the centre of world attention. On March 28, 1979, the Human Factor celebrated his greatest triumph to date...

A reactor at the Three Mile Island plant in Harrisburg, Pennsylvania, went haywire!

For a few dramatic days the plant leaked and there was a risk of a catastrophe. Tens of thousands of people fled... pregnant women were ordered out of the area...and children... while plans were made for the evacuation of a half-million people.

Three Mile Island was a close shave — so close that the US nuclear authorities said later they would have completely evacuated the area if they'd realised the true extent of the threat. The plant owners had been less than helpful in letting the authorities and the public know what was going on...

GOLDSBORO, PA: THE TOWN NEAREST
THE THREE MILE ISLAND NUCLEAR
POWER PLANT STANDS NEARLY
DESERTED EXCEPT FOR ONE MAN
RUNNING NEAR THE WATERS EDGE.

I MADE A COUPLE
OF WRONG MOVES

47

After Harrisburg, the nuclear industry could never again claim it was offering a safe product. Investigations of the accident revealed a long string of safety problems, as well as carelessness, misconduct and cover-ups.

Three Mile Island brought home to the public once and for all what nukes are really all about. It was a painful reminder that something distinguishes nuclear power from all other energy forms.

IN A WORD: RADIATION

DON'T MISS CHAPTER 4.

In which the China Syndrome is explained, reactor cores are cooled or caught, Rasmussen puts our minds at rest and the Third World is ripped off once again...

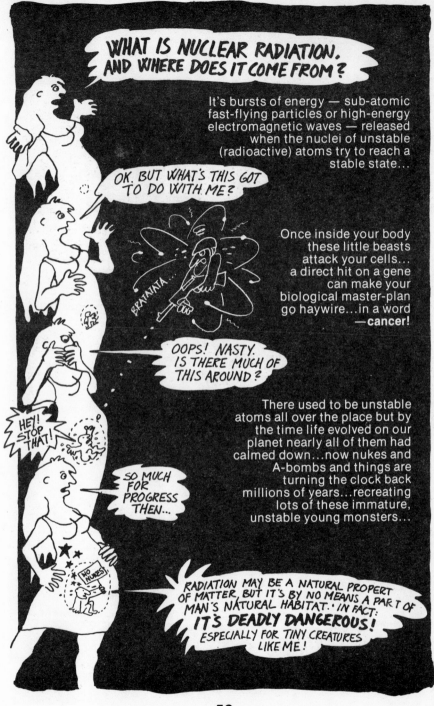

WHAT IS NUCLEAR RADIATION. AND WHERE DOES IT COME FROM?

It's bursts of energy — sub-atomic fast-flying particles or high-energy electromagnetic waves — released when the nuclei of unstable (radioactive) atoms try to reach a stable state...

OK. BUT WHAT'S THIS GOT TO DO WITH ME?

Once inside your body these little beasts attack your cells... a direct hit on a gene can make your biological master-plan go haywire...in a word —**cancer!**

BRATATATA...

OOPS! NASTY. IS THERE MUCH OF THIS AROUND?

There used to be unstable atoms all over the place but by the time life evolved on our planet nearly all of them had calmed down...now nukes and A-bombs and things are turning the clock back millions of years...recreating lots of these immature, unstable young monsters...

HEY! STOP THAT!

SO MUCH FOR PROGRESS THEN...

NO NUKES

RADIATION MAY BE A NATURAL PROPERT OF MATTER, BUT IT'S BY NO MEANS A PART OF MAN'S NATURAL HABITAT. IN FACT: **IT'S DEADLY DANGEROUS!** ESPECIALLY FOR TINY CREATURES LIKE ME!

50

Huge amounts of radioactive material are produced in the operation of nukes...the 'ashes' from the nuclear 'burning' of uranium fuel...and only a small leak can cause a catastrophe...

THE NUKE IS BOTTLED-UP RADIATION!

Were there any guarantee that it could be *kept* bottled up...after leaving the reactor as well...people wouldn't be making such a fuss...but there is no guarantee...

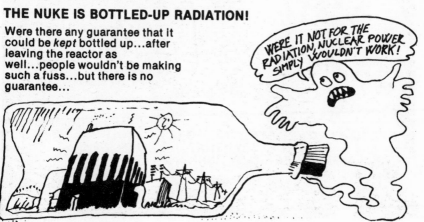

A TYPICAL NUCLEAR POWER REACTOR...

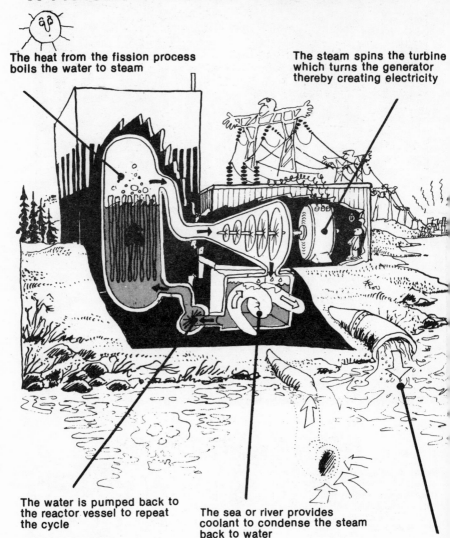

The heat from the fission process boils the water to steam

The steam spins the turbine which turns the generator thereby creating electricity

The water is pumped back to the reactor vessel to repeat the cycle

The sea or river provides coolant to condense the steam back to water

The discharged coolant is several degrees warmer

ORDINARY POWER STATIONS WORK IN THE SAME WAY EXCEPT THAT THE WATER IS HEATED BY BURNING OIL OR COAL...

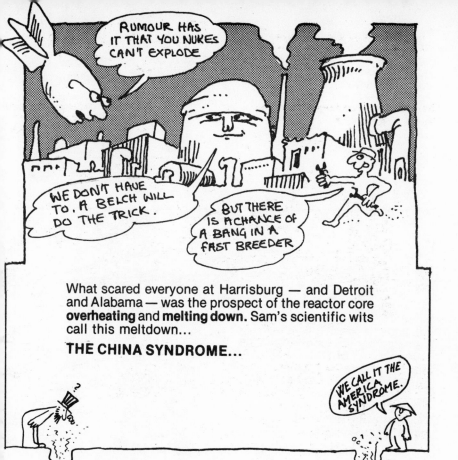

What scared everyone at Harrisburg — and Detroit and Alabama — was the prospect of the reactor core **overheating** and **melting down**. Sam's scientific wits call this meltdown...

THE CHINA SYNDROME...

NUCLEAR MELTDOWN

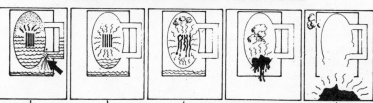

A main coolant pipe ruptures and the core temperature rises.

The emergency cooling system also fails and the fuel rods melt.

The fuel itself melts and the core fills with hot radioactive gases.

The molten fuel burns through the pressure vessel and containment dome.

Steam explosions fling metal parts through the dome walls...while the molten uranium burns down into the earth.

The most dangerous forms of radioactivity likely to be unleashed in a meltdown are the gaseous **fission products.** The exact consequences of such an accident — how many people would die how quickly — could only be established by a **test run**...

PERTH? FOREIGN OFFICE IN LONDON HERE... HOW MANY YOUR END?

A key barrier between a meltdown and a public health disaster is the ECCS — the emergency core cooling system. There's a lot of evidence that it's less than totally reliable. Special hearings in 1973 turned up two dozen safety researchers **from the US Atomic Energy Commission itself** with misgivings about the ECCS...

PUBLIC HEALTH

MELT-DOWN

Of course the nuclear industry is **fully confident** that the ECCS is a safe back-up...so confident that it's developing 'core catchers' in case the ECCS fails. These are designed to disperse and cool the molten debris. The French are installing one beneath their massive **Super-Phénix** breeder at Malville...

EEZ AN OLD FRENCH PROVERB: **EEF YOU CANNOT COOL ZEM, CATCH ZEM.**

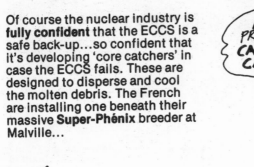

IMAGINE THAT THE WORST COMES TO THE WORST...Your local reactor is devastated by a major accident and a radioactive cloud drifts into your neighbourhood...

Those invisible particles aren't good for you so your best bet is to leave. They can settle into your food, drink, building materials or anything else that's lying around...**worst of all, though, is breathing them in**...

HOW MANY OF US WOULD BE KILLED IN A REACTOR DISASTER? ...AND WHAT ARE THE CHANCES OF IT HAPPENING?

For years lots of busy little men have been putting such questions to their slide-rules and computers. Depending on which report you trust it seems that between a dozen and tens of thousands of victims could be expected...

A test run would be useful to the statisticians...but it might be difficult to keep out of the newspapers...

NOT WHERE I COME FROM!

A major accident contaminating a large area apparently took place in the Soviet Union in 1958. Ivan hasn't let on what happened. It could have been a reactor breakdown but most signs point to an explosion of radioactive waste buried underground...

CLOSED FOR IDEOLOGICAL REASONS.

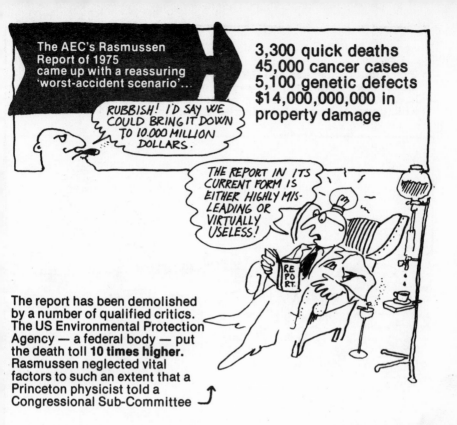

The AEC's Rasmussen Report of 1975 came up with a reassuring 'worst-accident scenario'...

3,300 quick deaths
45,000 cancer cases
5,100 genetic defects
$14,000,000,000 in property damage

RUBBISH! I'D SAY WE COULD BRING IT DOWN TO 10,000 MILLION DOLLARS.

THE REPORT IN ITS CURRENT FORM IS EITHER HIGHLY MIS-LEADING OR VIRTUALLY USELESS!

The report has been demolished by a number of qualified critics. The US Environmental Protection Agency — a federal body — put the death toll **10 times higher.** Rasmussen neglected vital factors to such an extent that a Princeton physicist told a Congressional Sub-Committee

Rasmussen's models were two plants in the US...where the law says no more than 380,000 people may live within 25 kilometres of a nuke. But elsewhere the situation can be very different...for instance, Sweden's Barsebäck plant is smack in the middle of Scandinavia's most densely-populated area...

By the way

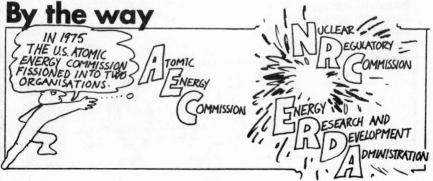

IN 1975 THE U.S. ATOMIC ENERGY COMMISSION FISSIONED INTO TWO ORGANISATIONS.

ATOMIC **E**NERGY **C**OMMISSION

NUCLEAR **R**EGULATORY **C**OMMISSION

ENERGY **R**ESEARCH AND **D**EVELOPMENT **A**DMINISTRATION

Over a million Swedes and Danes live within 25 kilometres of
Barsebäck and two million live within a 40 kilometre radius. The Danes
have no nukes themselves and are naturally thrilled at the Swedes'
willingness to involve them in the risks...

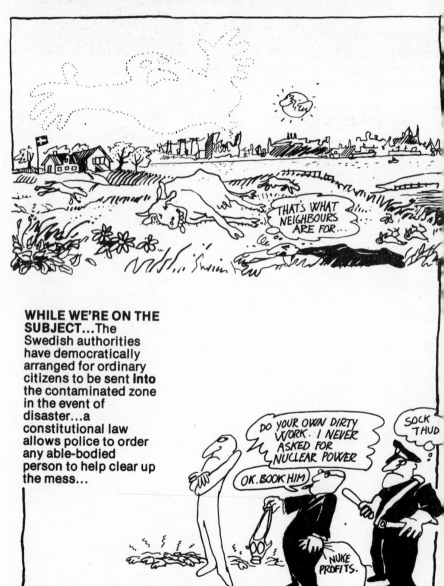

THAT'S WHAT NEIGHBOURS ARE FOR...

WHILE WE'RE ON THE SUBJECT... The Swedish authorities have democratically arranged for ordinary citizens to be sent **into** the contaminated zone in the event of disaster...a constitutional law allows police to order any able-bodied person to help clear up the mess...

DO YOUR OWN DIRTY WORK. I NEVER ASKED FOR NUCLEAR POWER

OK. BOOK HIM

SOCK THUD

NUKE PROFITS.

Rasmussen decided the chances of a disastrous accident occurring were extremely remote...

THERE WAS JUST AS MUCH CHANCE THAT A METEORITE WOULD HIT A U.S. CITY, HE SAID.

AND THE FOLLOWING DAY ONE DID?

No, but Browns Ferry and Harrisburg showed that risk assessment is a fool's game...

Moral: Rattlesnakes are harmless unless they bite.

RAZZLE DAZZLE

Just before Harrisburg, the NRC itself had attacked Rasmussen's sums, saying they were poorly supported...

NO SUMS MAKE SENSE WHEN YOU COUNT ME IN

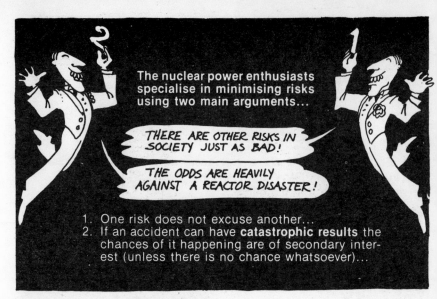

The nuclear power enthusiasts specialise in minimising risks using two main arguments...

THERE ARE OTHER RISKS IN SOCIETY JUST AS BAD!

THE ODDS ARE HEAVILY AGAINST A REACTOR DISASTER!

1. One risk does not excuse another...
2. If an accident can have **catastrophic results** the chances of it happening are of secondary interest (unless there is no chance whatsoever)...

WHY DO THE NUCLEAR AUTHORITIES KEEP INSISTING THAT THEIR NUKES ARE SAFE?

MY DEAR FELLOW, NO NUKES – NO NUCLEAR AUTHORITIES...

MOST REACTOR SAFETY PROGRAMMES IN THE WEST RELY ON THE FINDINGS OF THE AEC/NRC...ALTHOUGH IT HAS A LONG AND IMPRESSIVE RECORD OF MISLEADING THE PUBLIC ON THE SUBJECT...AND ON JUST ABOUT EVERY OTHER NUCLEAR MATTER...

After the special ECCS hearings a Congressional Committee said the AEC had **'developed a serious credibility gap...by suppressing unwelcome evidence of danger and by demoting or firing researchers who have pushed their findings too seriously...'**

The New York Times studied 11 years of AEC documents and concluded that the agency **'repeatedly sought to suppress studies by its own scientists that found nuclear reactors were more dangerous than officially acknowledged or that raised questions about reactor safety devices...'**

Leo Goodman, a union energy expert sometimes called the godfather of the anti-nuclear struggle in the US, put it more bluntly: **'AEC was dishonest and tricky from the word go...Watergate was nothing new to me...'**

FAR FROM BEING SAFE nukes have been besieged by problems from the outset…as pointed out in a 1973 AEC report which was suppressed until the environment group Friends of the Earth got hold of a copy…

The British among others have flatly rejected Sam's claims that his LWRs are safe…a programme to buy 32 of them was abandoned after the British government's Chief Scientific Advisor decided they could rupture without warning…

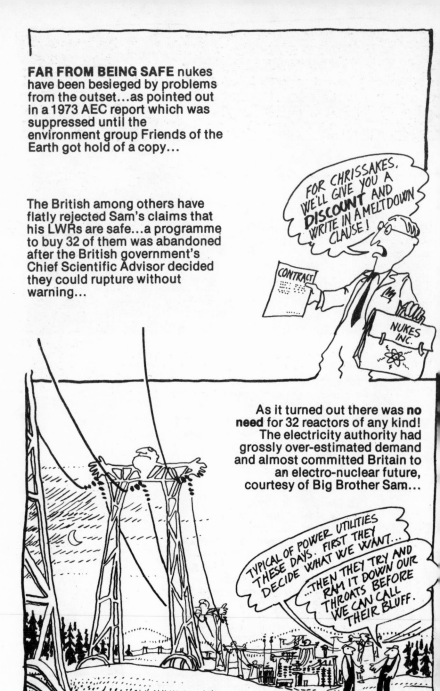

FOR CHRISSAKES, WE'LL GIVE YOU A **DISCOUNT** AND WRITE IN A MELTDOWN CLAUSE!

CONTRACT

NUKES INC.

As it turned out there was **no need** for 32 reactors of any kind! The electricity authority had grossly over-estimated demand and almost committed Britain to an electro-nuclear future, courtesy of Big Brother Sam…

TYPICAL OF POWER UTILITIES THESE DAYS. FIRST THEY DECIDE WHAT WE WANT…

…THEN THEY TRY AND RAM IT DOWN OUR THROATS BEFORE WE CAN CALL THEIR BLUFF.

Operating nukes on home ground is a risky enough exercise ...exporting them to non-industrial countries is a recipe for calamity...

Reactors are a bit more than a source of heat in a bucket of water...they're very complicated...and when fully built the **exported** nuke may be very different from the **original** model in the supplier state.

IN FACT, THERE MAY NOT EVEN BE AN ORIGINAL MODEL!

WOULD YOU MIND EX-PLAINING YOUR-SELF, YOUNG MAN.

The countries exporting to the Third World — France, West Germany, Canada and the US — mainly build big ones for the domestic market...nukes generating 1,000 Megawatts (MWe) of electricity or more...

MEGAWHAT?

MEGAWATT! THAT MEANS ONE MILLION WATTS OF ELECTRICITY!

BUT WE OF THE "LESS-DEVELOPED COUNTRIES" DON'T NEED GIGANTIC POWER PLANTS.

FOR SALE.

O.K. SO YOU GET A SITUATION LIKE THIS...

The true horror story

FROM I.A.E.A. BULLETIN 19 NO 2.

Egypt, South Korea and the Philippines buy a certain type of 600 MWe LWR from the US...

For safety reasons it's common practice for an exported nuke to be 'referenced to' — checked against — a similar one in the country of origin...

There's another problem...a reactor working like a dream on the banks of Lake Ontario may not be at all happy in Western Africa. Take the case of two West German LWRs in Iran...

MEIN GOTT! FRITZ, I THINK I'M GOING TO THROW UP.

They're being built in a relatively high seismic area but they're referenced to a plant in a low seismic area of West Germany

SO VOT?

SO WE MUST MAKE A LOT OF DESIGN CHANGES.

YES! FROM THE FOUNDATIONS...

...TO THE REACTOR COMPONENTS...

...AND THESE WILL NOT BE SUBJECT TO REVIEW BY THE WEST GERMAN SAFETY AUTHORITIES!

IAEA

BUT BECAUSE OF THE HOME DEMAND FOR **BIG** REACTORS, WE'VE NO CORRESPONDING MODEL FOR THEM TO REFER TO.

One being built in **Puerto Rico** was called off before completion...had it been finished it would have undergone systematic safety inspection and...

A NUMBER OF ALTERATIONS WOULD UNDOUBTEDLY HAVE BEEN MADE.

PROJECT ABANDONED

So instead the Egyptians, Koreans and Filipinos have to 'reference' their nuke to a similar one in Yugoslavia...which had been referenced to an earlier plant in Brazil...which in turn had been referenced to the abandoned Puerto Rican plant!

PUERTO RICO

BRAZIL

YUGOSLAVIA

WELL, AT LEAST THAT LEAVES US PLENTY OF SCOPE FOR IMPROVISATION.

Important changes and modifications made in domestic nukes in the exporting states are not necessarily included in the nukes flogged abroad...

FLOG FLOG ORIGINAL MODEL

THE MAIN VICTIMS, OF COURSE, ARE THE IMPORTING COUNTRIES LACKING SAFETY REVIEW BODIES OF THEIR OWN.

WHICH MEANS JUST ABOUT THE ENTIRE THIRD WORLD MAN!

THE
FAST
BREEDER
REACTORS

They're a bit different from today's burner reactors...

All nukes produce plutonium but only the breeders are designed to run on it...they generate more fissile material than they consume...and in theory they're eventually self-supporting...

It takes time for a breeder to produce enough plutonium to refuel another breeder...*not just years but decades!*

This 'doubling time' is put at 40 to 60 years for the French experimental plant in Marcoule — one of only a few operating breeders in the world today — but they're hoping for 30 to 40 years at the Super-Phénix in Malville...

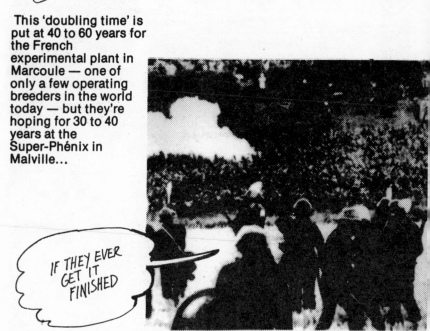

66

TO HAIL THE BREEDER AS THE SOLUTION TO THE COMING URANIUM CRISIS IS TO IGNORE THE DOUBLING TIME FACTOR

Given a 30-year doubling time Johnny would need *ten years* to get an extra truckload out of his breeder...not one year...assuming the thing worked at all...

By 1990 the IAEA estimates that 65-70 nukes will be commercially generating about 28-30,000 MWe in 17 **'less industrially developed countries'**...and there seems a good chance the first nuclear power disaster will occur in one of them...

The IAEA estimate for the **industrialised world** is over 500 nukes pumping out almost 400,000 MWe in 20 states by 1990...they've been at it longer...and as reactors are more accident-prone the older they get there's also a pretty good chance the first nuclear power disaster will occur in one of **these** countries...

YOU MEAN THERE'S NO REFUGE <u>ANYWHERE</u>?

TRY THE MOON...

COUNTRIES WITH COMMERCIAL NUKES
OPERATING, UNDER CONSTRUCTION OR PLANNED

1975 -	1980 -	1985 -	1990 -
Argentina	Austria	Iran	Cuba
Belgium	Brazil	Philippines	Egypt
Britain	Finland	Poland	Israel
Bulgaria	Hungary	Rumania	Thailand
Canada	South Korea	South Africa	Turkey
Czechoslovakia	Mexico		
France	Taiwan		
East Germany	Yugoslavia		
West Germany			
India			
Italy			
Japan			
Netherlands			
Pakistan			
Soviet Union			
Spain			
Sweden			
Switzerland			
United States			

STAND BY FOR
CHAPTER 5.

In which a cycle is not a cycle, radioactive waste is not disposed of, the spread of nuclear technology is not stopped and societies can not protect themselves from nuclear sabotage except by converting to a police state...

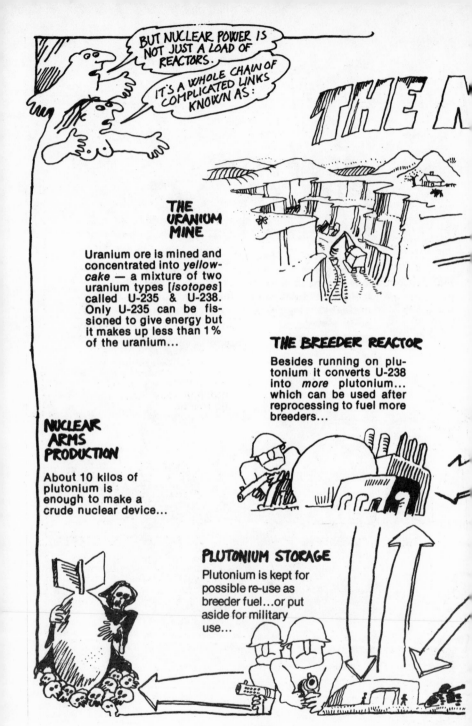

BUT NUCLEAR POWER IS NOT JUST A LOAD OF REACTORS.

IT'S A WHOLE CHAIN OF COMPLICATED LINKS KNOWN AS:

THE N

THE URANIUM MINE

Uranium ore is mined and concentrated into *yellow-cake* — a mixture of two uranium types [*isotopes*] called U-235 & U-238. Only U-235 can be fissioned to give energy but it makes up less than 1% of the uranium...

THE BREEDER REACTOR

Besides running on plutonium it converts U-238 into *more* plutonium... which can be used after reprocessing to fuel more breeders...

NUCLEAR ARMS PRODUCTION

About 10 kilos of plutonium is enough to make a crude nuclear device...

PLUTONIUM STORAGE

Plutonium is kept for possible re-use as breeder fuel...or put aside for military use...

UCLEAR FUEL CYCLE.

THE ENRICHMENT PLANT
By various techniques the U-235 content of the uranium is raised to 2-4%...

FUEL CONVERSION PLANT
The enriched uranium is converted into oxide pellets which are packed into fuel rods ready for the reactor...

THE REPROCESSING PLANT.
The spent fuel is broken down by acids...the unburnt uranium and plutonium are recovered...and the waste remains...and remains...and remains...

THE NUCLEAR REACTOR
The fuel fissions in the reactor core to generate heat for electricity production...and plutonium and radioactive 'fission product' waste...

WASTE STORAGE
Highly radioactive [*high-level*] waste is so toxic that it must be isolated from all life *for centuries or millenia!* Today it's stored while experts try and figure out how to get rid of it...

DANGER

73

A NUCLEAR FUEL CYCLE CAN BE DIFFERENT THINGS TO DIFFERENT PEOPLE...

AND THIS IS WHAT THE LONDON FINANCIAL TIMES WOULD HAVE US BELIEVE...

NO UNPLEASANT LEFT-OVERS LIKE RADIOACTIVE WASTE OR HOARDS OF PLUTONIUM...

IT JUST GOES ROUND AND ROUND AND ROUND...

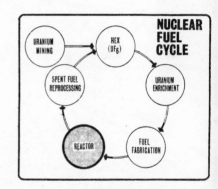

In reality it's more like a game of snakes and ladders played out in a minefield. All along the way there are hazards to public health...but it's the worker in the nuclear industry who's the most directly threatened... especially in the mines, in reactor repairs and waste 'management'...

COMPLAINTS, ALWAYS COMPLAINTS. YOU GET YOUR WORKING CLOTHES FREE, DON'T YOU.

The left-overs after uranium is extracted from ore are called *tailings* and they're radioactive...which came as a bit of a shock to people in the US southern states who had found them lying about in sand-like heaps and decided they'd make fine foundation and building materal...

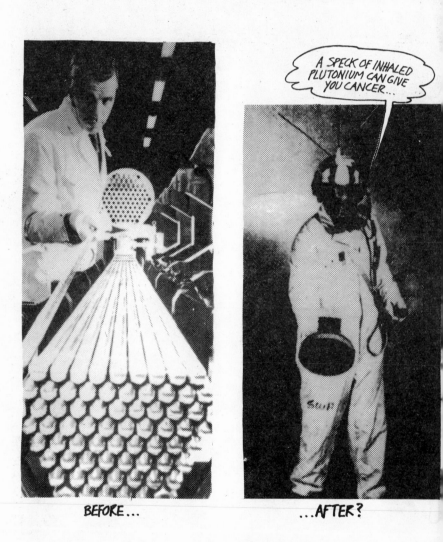

BEFORE...   ...AFTER?

SAFETY AT ENRICHMENT AND FUEL FABRICATION PLANTS IS NOT A BIG PROBLEM because the uranium fuel is still in a fairly tame state...*but the large-scale use of plutonium as fuel would change the picture completely...*

THE ENRICHMENT PLANTS ARE HUGE...

YOU MAY NOT BE SURPRISED TO HEAR THAT I'M THE MARKET LEADER.

They cover hundreds of acres ...they're tremendously expensive...and they consume as much electricity as a city of half a million people...

The *Russians* and *Chinese* have their own, in Siberia and Lanchow. The European Common Market countries have ganged up to build commercial plants in *Britain* (Capenhurst), *France* (Tricastin) and the *Netherlands* (Almelo)...

ALSO, OF COURSE, URANIUM ENRICHMENT CAN PROVIDE THE RAW MATERIAL FOR A-BOMBS...AS THE PEOPLE OF HIROSHIMA LEARNED IN 1945...SO DESPITE THE COST ENRICHMENT FACILITIES ARE QUITE A POPULAR ITEM IN THE NUCLEAR CATALOGUE...

ENRICH THE RICH, BOMB THE POOR...

IRAN

THE WEST GERMANS ARE MAKING ONE SPECIALLY FOR ME.

AND THEY SOLD THE OLD SHAH A SLICE OF THE ACTION

SLURP SLURP

BRAZIL

Also in the race are **Canada** (James Bay), **Japan** (Tokai Mura) and **South Africa** (Valindaba). A test plant began operating at Valindaba in 1975...which may explain why South Africa were said in 1977 to be in a position to test a nuclear bomb.

77

AS FOR THE REACTOR...

The nuclear industry is always issuing publicity photos like this to show us how nice and safe it is to work in a nuke...

For some reason photos of repairmen in action in a real emergency don't find their way into the PR brochures... and very rarely into the newspapers...

MAYBE PHOTOGRAPHERS AREN'T TOO KEEN ON RISKING THEIR LIVES...

The Swiss nuclear power industry once ran out of top-flight welders! All 700 who were qualified for the job collected the maximum radiation dose when repairing a reactor at Beznau...radiation was so intense that each welder could work for only two minutes...

On a breakdown at the Indian Point reactor outside New York *2,000 repairmen* were used in the same way...

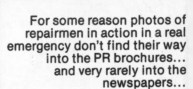

HI MAN. HOW ARE THINGS?

GREAT! JUST GOT TWO MINUTES WORK AT WINDSCALE.

LABOUR EXCHANGE

Top: THE OFFICIAL PICTURE…
BLUE SKIES & UNSPOILED NATURE…
AND IT'S SO HARMLESS YOU CAN
LIVE RIGHT NEXT DOOR…

Middle: THE REACTOR AS IT MIGHT
BE SEEN BY THE OPPONENT OF
NUCLEAR POWER WHO JOINS A
DEMONSTRATION…

Right: THE HOUSEHOLD REACTOR
(MID-21st CENTURY) AFTER BEING
TAKEN OUTSIDE BY THE REPAIRMAN…

79

OVER AT THE RE-PROCESSING PLANT

There's a lot of plutonium and radioactive fission products being handled ...so almost everything has to be operated by remote control from behind thick shielding...

The French plant at La Hague was the only one by 1978 capable of treating LWR oxide fuel...a lot of countries were queuing up...and as the order-book swelled the management exposed the workers to increasing doses of radiation. In the face of union protests more and more unorganised, short-time labour was employed...often with only a couple of days' training!

TROUBLE IS, CLEAN-UP AND REPAIR WORK SOMETIMES REQUIRES HUMAN BEINGS...

La Hague workers told union investigators...

OFTEN NO CONTAMINATION CHECKS WERE MADE AT THE EXITS...

I'VE SEEN WORKERS GO HOME EVEN THOUGH THE DETECTION METERS WENT BERSERK AS THEY APPROACHED...

I SAW A FELLOW WITH BREATHING TROUBLE LIFT HIS GAS-MASK IN A RADIO-ACTIVE ZONE TO GET A BIT OF "FRESH AIR"...

The workers at La Hague had to strike for months before the French nuclear industry would do anything about their working conditions...

THINGS START TO GET REALLY MESSY

A reprocessing accident can be at least as dangerous to the public as a reactor accident...yet the reprocessing plant lacks most of the reactor's safety barriers...

Radioactivity could escape into the environment if the witches' brew leaked out of the acid pond...or the coolant failed and the waste boiled...or the ventilation packed up and there was a chemical explosion...

REPROCESSING IS SUCH A DIFFICULT AND RISKY BUSINESS THAT IT HAS MORE OR LESS BEEN ABANDONED IN THE U.S....WITH GIANT LOSSES. GETTY OIL, GENERAL ELECTRIC, GULF AND SHELL HAVE ALL BURNED THEIR FINGERS...

Some people haven't given up, though. New plant for reprocessing oxide fuels is planned or being built in Britain (Windscale), West Germany (Karlsruhe), India (Tarapur) and Japan (Tokai Mura)...and La Hague aims to expand...

Finding a final resting place for high-level nuclear waste is a problem that has baffled scientists ever since the stuff started piling up in the 1940s...

Interim storage is a neat way of passing the buck...but liquid radioactive waste is a sizzling peril. In the United States more than 400,000 gallons have seeped away from one storage facility at Hanford on the Columbia River...

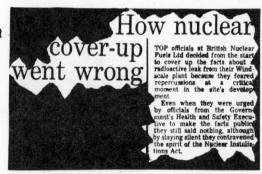

There was also a big leak from Windscale's storage tanks in 1976...

The Windscale management reacted in the true spirit of the nuclear establishment, which has always had the welfare of the general public at heart...

How nuclear cover-up went wrong

TOP officials at British Nuclear Fuels Ltd decided from the start to cover up the facts about a radioactive leak from their Windscale plant because they feared repercussions at a critical moment in the site's development.

Even when they were urged by officials from the Government's Health and Safety Executive to make the facts public they still said nothing, although by staying silent they contravened the spirit of the Nuclear Installations Act.

ALL PROPOSALS FOR LONG-TERM STORAGE OR DISPOSAL OF HIGH-LEVEL WASTE FROM THE NUCLEAR POWER INDUSTRY LIE AT THE RESEARCH AND DEVELOPMENT STATE...WHILE STOCKPILES GROW FASTER AND FASTER AROUND THE GLOBE...

There have been lots of *theoretical* solutions...

But in the words of the Nobel Prize-winning physicist Hannes Alfvén: *If a problem is too difficult to solve one can not claim that it's solved by pointing to all the efforts made to solve it...'*

BECAUSE OF THE IMMENSE TIME IT TAKES FOR HIGH-LEVEL WASTE TO COOL DOWN AND BECOME HARMLESS NO METHOD CAN BE GUARANTEED COMPLETELY SAFE...

66 *We must assume that these wastes will remain dangerous and will need to be isolated from the biosphere for hundreds of thousands of years. In considering arrangements for dealing safely with such wastes man is faced with time scales that transcend his experience...* 99

— Flowers Report to British Parliament September 1976

83

Exhaustive research has failed to turn up anything but a couple of reasonably promising ways of solidifying the waste...and the bright idea that it would be best stored where it's retrievable in case things go wrong...

CAN'T YOU LEAVE THE SPENT FUEL INSIDE THE RODS INSTEAD OF REPROCESSING IT?

THERE'D BE FEWER PROBLEMS... BUT IT WOULD STILL REQUIRE ULTIMATE DISPOSAL.

AND IT WOULD MAKE ME REDUNDANT...

How long it takes for high-level waste to become docile depends on how long it takes for its ingredients to decay. Most of the *fission products* have calmed down after 800 years...but other radioactive elements called *actinides* survive longer, including traces of plutonium not stripped out of the fuel in reprocessing...

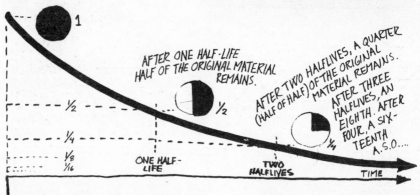

AFTER ONE HALF-LIFE HALF OF THE ORIGINAL MATERIAL REMAINS.

AFTER TWO HALFLIVES, A QUARTER (HALF OF HALF) OF THE ORIGINAL MATERIAL REMAINS. AFTER THREE HALFLIVES, AN EIGHTH. AFTER FOUR, A SIXTEENTH A.S.O....

ONE HALF-LIFE

TWO HALFLIVES

TIME

The rate-of-decay of a waste product is measured in its **half-life** — the time it takes for half the amount of material to disintegrate. For example Caesium-137 has a half-life of 30 years...so reducing radioactivity a thousand-fold takes 300 years (ten half-lives). Species like Strontium 90 **(28 years)**, Curium-244 **(18 years)** and Caesium-137 are more dangerous than long-life ones like Plutonium-239 **(24,400 years)** because their radioactivity is more concentrated...

THE PROBLEM IS THAT HEAVY, RADIOACTIVE ELEMENTS LIKE PLUTONIUM DECAY INTO OTHER RADIOACTIVE ELEMENTS, WHICH DECAY INTO OTHERS AND SO ON...AND DON'T FORGET THAT, APART FROM THEIR RADIATION, THESE ELEMENTS ARE HIGHLY POISONOUS!

High-level waste is *not the only problem*...there are *medium-activity* and *low-activity* left-overs as well...in much larger quantities...

Most medium-level waste is toxic enough to require isolation for centuries. It can include anything from protective clothing to old bits of reactor...

Low-level waste is routinely discharged into the surroundings, buried in the ground or dumped at sea...

DISUSED REACTOR- AND REPROCESSING PLANTS ARE ALSO A FORM OF WASTE... RADIOACTIVE RUINS THAT WILL STAND AS MONUMENTS TO... WELL... ER... SOMETHING...

CLICK

WHERE DOES THE BREEDER COME IN?

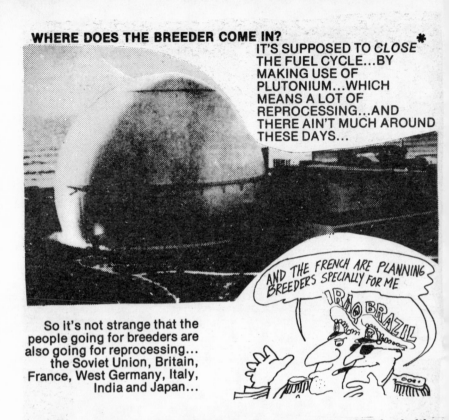

IT'S SUPPOSED TO *CLOSE* THE FUEL CYCLE...BY MAKING USE OF PLUTONIUM...WHICH MEANS A LOT OF REPROCESSING...AND THERE AIN'T MUCH AROUND THESE DAYS...

AND THE FRENCH ARE PLANNING BREEDERS SPECIALLY FOR ME

IRAQ BRAZIL

So it's not strange that the people going for breeders are also going for reprocessing... the Soviet Union, Britain, France, West Germany, Italy, India and Japan...

President Carter set the cat among the pigeons in 1977 by halting commercial reprocessing and plutonium recycling in the US. But his appeal for others to follow suit was met with suspicion...

HA! IF YOU ARE SERIOUS, WHY DON'T YOU CANCEL YOUR **CLINCH RIVER** BREEDER?

YOU JUST WANT TIME TO CATCH UP, N'EST-CE PAS?

ZU LATE! ME UND MEIN FREUND ARE TAKING OVER YOUR MARKETS, UNDERSTAND?

BREEDERS FOR SALE

* The only way to close the fuel cycle is to shut your eyes and pretend there's no such thing as radioactive waste...

86

THE FUEL CYCLE STRETCHES ROUND THE WORLD...

I'M NOTHING IF NOT INTERNATIONAL!

Japan may buy uranium from South Africa, convert it in Britain for enrichment in the US, push it through a reactor at home, send the spent fuel to France for reprocessing and get back a load of plutonium...

PLUS A LOAD OF HIGH LEVEL WASTE.

THAT'S OK. WE'LL BURY IT IN SOUTH KOREA.

Many think the risk of accident, theft or sabotage is greatest during the transport of nuclear materials around the fuel cycle...

They're moved by road, rail, sea and air...and today's flow of traffic promises to swell to a flood over the next two decades...

ACCIDENTS HAPPEN!
SHIPS SINK!
TRUCKS CRASH...
SO DO TRAINS & PLANES!

PEOPLE GET CARELESS!
PEOPLE GET CORRUPTED!
PEOPLE GET BLACKMAILED!
THIEVES STEAL!

TERRORISTS
GO IN
SHOOTING!

HOLD ON... WE TAKE THE UTMOST PRECAUTIONS WITH NUCLEAR FREIGHT

SURE! UNTIL THEY GET TOO EXPENSIVE OR SLOW YOU DOWN.

THE NUCLEAR PARK SOLUTION...

The industry claims that the latest reactors are built to withstand a crashing plane. But not a Jumbo-size airliner. Nor a big bomb. As for reprocessing plants, they're almost totally unprotected from things falling out of the sky...

IN THE EVENT OF WAR...NUCLEAR POWER IS A LIABILITY...

Nukes are such obvious and potentially disastrous targets that the only sensible thing to do if hostilities break out is to *close them down*...so countries with most of their energy eggs in the nuclear basket will be fighting the war in the dark...

A TRIP DOWN PARANOIA GULCH...

What people usually mean by the 'wrong hands' are those belonging to guerilla groups, organised crime syndicates, stray madmen or leaders of the Third World...
as opposed to characters of
infinitely sound judgement
like Richard Nixon, Moshe
Dayan, NATO and Warsaw Pact
generals...

The name of the game is...
CLANDESTINE DIVERSION OF PLUTONIUM
or...
**NUCLEAR PROLIFERATION AT GOVERNMENTAL
AND NON-GOVERNMENTAL LEVEL**
or quite simply...
SPREADING THE BOMBS AROUND

Two years after the Indian Bang showed *Atoms for Peace* and *Atoms for War* to be Siamese Twins the record was officially put straight by an NRC Commissioner:

'International action to control (the dangers of proliferation) associated with the civilian nuclear fuel cycle depends critically on the understanding of two facts. First, that nuclear weapons can be manufactured from reactor-grade plutonium. And second, that for any nation that has done its homework, separated plutonium can be suddenly appropriated from its storage place and inserted in warheads within days.'

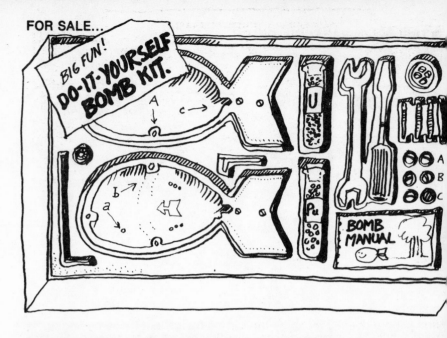

You don't need a big nuclear power programme to make nuclear weapons under the counter...as the Indians demonstrated. A small research reactor...laboratory-scale reprocessing...and the whole thing was said to cost only $400,000...

A humble little 40MWe reactor gives enough plutonium a year for three Hiroshima-size bombs. The Stockholm International Peace Research Institute (SIPRI) says the components can be discreetly bought on the open market for $20 million...and small reprocessing and uranium enrichment units are also available to those anxious for bomb materials...

THE NEW SUPERSALESMEN...

THE SECRET OF OUR SUCCESS IS THAT WE'RE UNPREJUDICED — WE'LL DEAL WITH ANYONE.

JA, RACISTS, FASCISTS, MILITARY DICTATORS, RELIGIOUS FANATICS, YOU NAME IT.

YOU CAN SAY THAT AGAIN.

GOD BLESS THE FREE MARKET

SOUTH AFRICA

LIBYA

France and West Germany won a place in every pacifist's heart by contracting to sell the direct route to nuclear bombs — missing links in the fuel cycle — to Pakistan and Brazil respectively.

Today, in the shadow of the Soviet-American nuclear arms race, local rivals are engaged in contests of their own. The Pakistanis are scrambling to catch up with the Indians...the Brazilians are vying with Argentina for nuclear leadership of Latin America...and both Egypt and Israel are already believed to have access to nuclear weapons.

NONE OF THESE NATIONS ARE A PARTY TO THE NON-PROLIFERA-TION TREATY [NPT] WHICH WAS SUPPOSED TO STOP THE SPREAD OF NUCLEAR ARMS...

DOWN HERE IN SOUTH AFRICA WE NEED NO ENCOURAGE-MENT TO DEVELOP A NUCLEAR THREAT...

WHO'S AFRAID OF THE **N.P.T.** ?

A lot of non-nuclear countries, for a start...They see it as a neat way for the big powers to hog the nuclear hardware. And they have some cause for resentment...The British, Americans and Russians have consistently refused to promise they won't use their nuclear arms against non-nuclear states — even the ones that sign the NPT!

HOWEVER... WE MIGHT BE PREPARED TO NEGOTIATE THE **NUMBER OF WARHEADS** WE'LL USE ON YOU...

DA!

OUR GRATITUDE KNOWS NO BOUNDS

NON-NUCLEAR STATES

If you read the small print it's very clear why many countries choose not to sign the NPT...*the treaty obliges signatories to help non-signatories with nuclear energy programmes wherever possible!*

WASHINGTON?... WE DEMAND OUR RIGHTS UNDER THE **NPT.** SEND A COUPLE OF REACTORS, ONE OF THOSE REPROCESSING THINGS, AND A SIX-PACK OF URANIUM 235.

CASH ON DELIVERY?

THAT'S HOW INDIA GOT HOLD OF THAT CANADIAN REACTOR...IT JUST WENT AND ORDERED IT!

92

FROM ONE PAPER TIGER TO ANOTHER...

Even when it's signed the NPT is not much of a deterrent as membership can be cancelled at 3 months' notice...or simply ignored in a 'national emergency'...then you just clap a warhead onto the missile the NPT allows you to assemble and off you go...

Many experts doubt that legal safeguards are enforceable anyway... Treaties invariably have loopholes...nuclear inventories are quite easy to fiddle...bomb tests can be simulated on computers...and an inspector's task can be made virtually impossible on site...

NO, INSPECTOR, IT'S AN INDOOR TENNIS COURT...

KEEP OUT

WARNING

So the nuclear 'haves' decided to lean on the 'have-nots' a bit more heavily...they formed THE LONDON CLUB...

A HUSH-HUSH GATHERING OF 14 NUCLEAR EXPORTERS – IN OTHER WORDS, MOST OF THE INDUSTRIALIZED NATIONS.

The London Club claims i... wants to stop the spread o... sensitive technology and know-how to the 'wrong quarters'. At the same time, its members are competing with one another in a cut-throat multi-billion dollar export industry...

AND WE ALL KNOW THE UPSHOT OF CONFLICTS LIKE THIS.

WHO'S WILLING TO LOSE MARKETS JUST FOR THE SAKE OF WORLD PEACE?

so it's no surprise that they can't agree which countries to sell their goodies to under what circumstances...or rather which countries not to sell their goodies to...

GUESS WHO'S MAKING MOST OBJECTIONS?

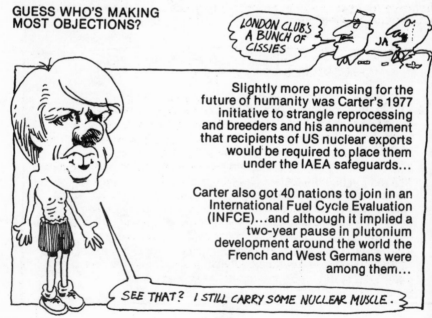

LONDON CLUB'S A BUNCH OF CISSIES

Slightly more promising for the future of humanity was Carter's 1977 initiative to strangle reprocessing and breeders and his announcement that recipients of US nuclear exports would be required to place them under the IAEA safeguards...

Carter also got 40 nations to join in an International Fuel Cycle Evaluation (INFCE)...and although it implied a two-year pause in plutonium development around the world the French and West Germans were among them...

SEE THAT? I STILL CARRY SOME NUCLEAR MUSCLE.

94

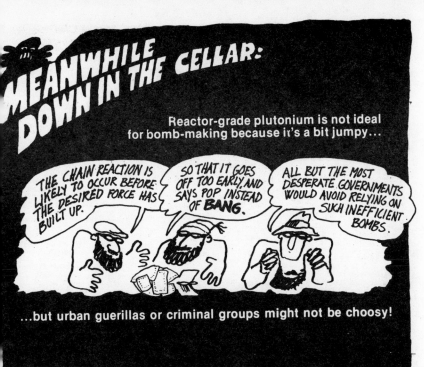

MEANWHILE DOWN IN THE CELLAR:

Reactor-grade plutonium is not ideal for bomb-making because it's a bit jumpy...

THE CHAIN REACTION IS LIKELY TO OCCUR BEFORE THE DESIRED FORCE HAS BUILT UP.

SO THAT IT GOES OFF TOO EARLY AND SAYS POP INSTEAD OF **BANG**.

ALL BUT THE MOST DESPERATE GOVERNMENTS WOULD AVOID RELYING ON SUCH INEFFICIENT BOMBS.

...but urban guerillas or criminal groups might not be choosy!

Authorities like the NRC say reactor-grade plutonium can be used by such outsiders *on the basis only of existing information in the open literature* to make crude but convincing bombs...of unpredictable yield but enough to wipe out a seat of government...

SIPRI SUMMED UP THE HAZARDS LIKE THIS...

66 Plutonium could be stolen and sold on a «black market», or it could be ransomed, or a group of criminals might steal plutonium for profit, or for use as a nuclear threat to deter police or otherwise further their activities. Or one of the more than 50 terrorist groups that are said to exist worldwide might see nuclear weapons as means of enhancing its capability to use, or threaten, violence. Or a revolutionary-minded political group within a country might aquire nuclear weapons to achieve its political objectives or to deter violence against it. And it should not be forgotten that the danger inherent in a crude nuclear device constructed by, for example, a terrorist group is not confined to a possible nuclear explosion. The contamination of a large area by high levels of plutonium would be an enormous threat in itself. 99

A London newspaper which received a 'credible' A-bomb design from a group of students asked some scientific and weapons experts to describe in concrete terms the amateur nuclear weapon threat...

THE TERRORIST A-bomb is about the size of a tea-chest, weighs around half a ton, could be conveniently carried in an inconspicuous vehicle such as a Transit van, and would be detonated remotely by radio.

If the yield of one kiloton was achieved its fireball would be about 70 metres in diameter, vapourising much of the area where it touched the ground. All exposed people within a radius of half a kilometre would be killed directly by thermal and other radiation, while devastation of structures by blast might extend over a radius of about 1 kilometre. In some meteorological conditions, such as a surface temperature inversion which is common in Britain at dusk and during the night, blast damage might extend over a radius of up to 3 kilometres.

Because this would be a surface burst the radioactive plume would be large and would contain a great deal of particulate material which, re-condensed from the fireball, would contain highly radioactive fission products and core residues. As in other nuclear incidents the behaviour of this cloud of dangerous material would depend upon meteorological conditions, but airborne radioactivity is impossible to control and would descend on random segments of the population before any

effective warning could be given. Among those affected and heavily exposed death might ensue within seven weeks. A much larger segment of the population might be affected in a way which, much later in life, produced cancer in some form.

If such a weapon were exploded in, say, the car park outside the House of Lords, the seat of Government might be destroyed and rendered ineffective for a long period. The cost of clean-up alone, on present industrial estimates of about £1 million per gramme of plutonium, would be well over £1 billion. The cost of damage repair is inestimable but would probably be of the same order.

* Premature detonation means the thing might go off a millionth of a second too early — not while it's being built or transported...

* Safe handling of strategic materials and high explosives is widely known and described in many textbooks...

* Chemical conversion of the materials is not much riskier than heroin synthesis...and that's been going on in criminal laboratories for a long time...

96

GETTING HOLD OF THE PLUTONIUM MAY BE ONE OF THE EASIER PARTS OF THE OPERATION…THE HISTORY OF PLUTONIUM MANAGEMENT IS FULL OF UNEXPLAINED LOSSES, CARELESSNESS AND 'COOKED BOOKS'…

Accounting systems for nuclear materials can never be more than 98-99% accurate…and 1% of the annual throughput at a reprocessing plant, for example, is enough for a cartload of bombs…

Thousands of people are handling plutonium every day in the nuclear industry. There's 'statistical uncertainty' about *4 or 5 tons* of potential nuclear weapons material in the US alone — say about 1,000 bombs. Perhaps none of it has been taken…or all of it! As no-one knows for sure, all threats that sound technically feasible have to be taken seriously…

WELCOME TO THE POLICE STATE...

SIPRI: 'Societies can not effectively protect the nuclear fuel cycle against sabotage short of converting to a garrison state'

In the security business it's generally agreed that you can't protec anything from people determined enough to get it...if that were possible there'd be no plane hijacks or bank robberies...And here again we have to take into account the human factor...

IF THE NUCLEAR POWER INDUSTRY'S INSISTENCE THAT IT CAN SAFELY GUARD THE FUEL CYCLE IS TO BE TAKEN SERIOUSLY SOME DRASTIC STEPS WILL BE NEEDED VERY SOON...

HUMAN FACTOR: DIG THIS: EVERY YEAR ABOUT 3% OF THE 120,000 PEOPLE WORKING WITH U.S. NUCLEAR WEAPONS ARE RELIEVED OF DUTY BECAUSE OF DRUG USE, MENTAL INSTABILITY OR SOME OTHER RISK.

All dissent in a Fissile Society can be viewed as a threat...so the least we can expect is an escalation of police & military surveillance...in fact what measures could not be justified in the name of NUCLEAR SECURITY?

...AND WE ARE GOING TO KEEP YOU ALL INDOORS UNTIL YOU GIVE US BACK THAT TRUCKLOAD OF PLUTONIUM!

THE WRITING'S ALREADY ON THE WALL...

Britain, we're told, is a stronghold of democratic tradition...The *Atomic Energy Authority [Special Constables] Act* of 1976, which legalised an armed nuclear police force, showed that plutonium security and democratic controls don't mix...

This little army of atomic cops *is not directly accountable to a government minister*, only to the AEA. One study has suggested that this is because it's recognised in the corridors of power that some nasty breaches of civil liberties might be needed in a plutonium emergency. In which case the government might find it convenient to plead ignorance...

In the US at least one private nuclear police force (VEPCO's) is seeking the same powers of pursuit, arrest and access to confidential files as the regular state police. And there's talk on Capitol Hill of creating a special federal force that can skip bothersome details like court orders and formal charges, and even under certain circumstances torture suspected 'nuclear terrorists'...

99

NO NUKES

NO!

In many parts of the world — especially Western Europe and North America — the general public is growing more and more disturbed about nuclear power. The authorities can no longer proceed unchallenged. The public is demanding inquiries, fighting building applications and staging giant rallies and demonstrations in numerous countries...

A situation has arisen in which public participation in energy questions is seriously disrupting governments' plans for nuclear expansion...

Pushing on down the nuclear path in the face of public opposition is as good a way as any of sparking *civil disobedience*. Direct action has already begun in many countries. For example there've been...

.. SO I'M AFRAID WE HAVE TO CHANGE THE GROUNDRULES. AS FROM NOW PUBLIC INVOLVEMENT IN OUR AFFAIRS IS **ILLEGAL!**

OCCUPATIONS of proposed nuclear sites in the US, France, Britain and West Germany...

BOMB ATTACKS on nuclear installations in France, Spain, Switzerland and the US...

LEAKS of stolen confidential nuclear information in Australia.

If governments remain unswayed direct action against fuel transports and existing fuel cycle facilities also seems likely as well as interference with national electricity supplies...

AS YOU STILL WON'T ACCEPT THAT WE KNOW BEST WE'RE CONVERTING TO A **POLICE STATE**

100

A TECHNOLOGY THAT SOCIETY MUST ADAPT TO...

A plutonium economy demands centralised, authoritarian rule...with fewer people wielding more power...and erosion of civil rights...starting with the right to organise labour...

HERE IN AUSTRALIA THEY HAVE ALREADY OUTLAWED INDUSTRIAL ACTION IN THE NUCLEAR INDUSTRY.

WORKERS UNITE

KANGA-ROONI-ON

Add the political and social risks to the proliferation and safety hazards of nuclear power and the picture begins to emerge of a highly oppressive technology...

WHO NEEDS NUKES?

The public?

NO THANKS...

The Scientists?

NO, THEIR TALENTS CAN BE BETTER EMPLOYED ELSEWHERE...

The Military?

YES! IT JUSTIFIES THEIR EXISTENCE!

The Engineers?

NO, FOR THE SAME REASON...

State Governments?

SOME OF THEM HAVE STAKED THEIR FUTURE ON IT. (AND OURS...)

IS IT PROFITABLE?

Monopoly Capitalists?

ONLY AS LONG AS IT IS PROFITABLE... OR IT UNDERPINS THEIR PROFITS IN OTHER SECTORS... OR PRESERVES THE POLITICAL STATUS QUO SO THAT THEY'RE FREE TO ACT...

If governments weren't propping up the nuclear industry it would soon go under. Quick profits can still be made here and there...but on the whole it's a bad investment in straight capitalist terms.

PUBLIC RESISTANCE is one of the main reasons that nuclear electricity, once hailed as a bargain, now looks a dubious economic bet. In the US, delays and safety modifications forced through by environment groups helped send *capital costs* rocketing during the 1970s...

SO FUEL PRICES ARE NOW OF SECONDARY INTEREST... AND WE BREEDERS ARE LESS ATTRACTIVE THAN EVER!

Capital costs of conventional power stations are increasing, too, but not so fast. US analysts expect the price of nuclear electricity to catch up with the price of coal-based electricity around the mid-1980s. One of them, Barry Commoner, thinks nukes will soak up so much money that they'll lead to *capital starvation*.

IN THE END, HOW LONG THE LAME NUCLEAR DUCK IS KEPT AFLOAT MAY DEPEND ON HOW BADLY GOVERNMENTS WANT TO GET FREE OF OPEC AND ALL THOSE MILITANT COAL-MINERS....

LOOKS LIKE THE ONLY WAY WE'RE GOING TO OFF THE NUKE IS BY **POLITICAL MASS ACTION**

RIGHT. POPULAR MOVEMENTS LINKING **ALL** NUCLEAR OPPOSITION.

BUT WE MUST HAVE A CLEAR UNDERSTANDING OF WHAT WE'RE **AGAINST** AND WHAT WE'RE **FOR!**

PEOPLE'S POWER!

TUNE IN TO
CHAPTER 6.

In which we're invited to choose between hard & soft, the Energy Junkie tries to justify his habit, Illich and Lovins have a say and some Third World countries take matters into their own hands...

SAYING 'NO' TO NUKES
MEANS SAYING 'YES' TO SOMETHING ELSE!

Little would be won if it meant saying 'yes' to ferocious consumption of coal, oil & gas...their risks are well-known...

* As a workplace the coal mine is as hazardous as the uranium mine...

* The death rate for oil industry divers in the North Sea is notoriously high...

* Tankers carrying frozen natural gas can explode like a small A-bomb...at sea or in harbour...

Then there's the threat to the natural environment...

Oil tankers keep colliding, sinking and running aground, often with disastrous consequences...and oil rigs blow out or leak...and «acid rain» from the burning of oil and coal pollutes lakes and soil...sometimes very far afield...

HERE IN NORWAY WE'D LIKE TO THANK BRITISH INDUSTRY FOR KILLING OUR FRESHWATER FISH.

Also, burning coal, gas & oil raises the carbon dioxide content in the Earth's atmosphere...heating it up...so continued use of fossil fuels at today's rapid rate can upset the world climate...

UT COMPARING THE RISKS IS MISSING THE POINT.

Must we choose between cholera and the plague…

…or can we instead go for a clean bill of health?

Technological development and social development go hand in hand. Today's decisions in the energy field are going to shape lifestyles into the next century…and the effects could be hard to reverse…

HERE'S ANOTHER KIND OF CHOICE…

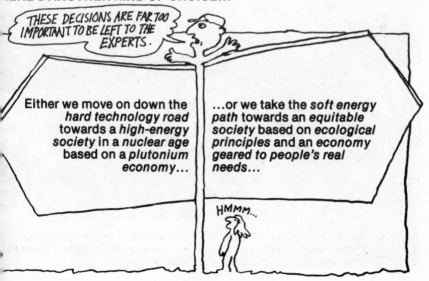

THESE DECISIONS ARE FAR TOO IMPORTANT TO BE LEFT TO THE EXPERTS.

Either we move on down the *hard technology road* towards a *high-energy society* in a *nuclear age* based on a *plutonium economy*…

…or we take the *soft energy path* towards an *equitable society* based on *ecological principles* and an *economy geared to people's real needs*…

HMMM…

CAPITALISTS ASK A DIFFERENT SET OF QUESTIONS...

Thanks to this line of thinking we're in the grip of a system that's using up the Earth's non-renewable resources ...destroying the natural environment...reducing work to monotonous and meaningless drudgery...producing a lot of rubbish while masses of people starve around the world...

The official line in the industrialised countries is that demand for energy will continue to soar...and that around 1990 it'll outstrip supplies of fossil fuels...creating an ENERGY GAP that only the nuke can fill...

IN FACT, THEY SAY, WITHOUT NUCLEAR ELECTRICITY OUR LIVES WILL BE TURNED UPSIDE DOWN BY THE END OF THE CENTURY...

CAPITALISM HAD ITS POINTS...

IT'S THE SYSTEM THAT'S GROWING. NOT THE QUALITY OF LIFE!

The capitalist market economy has brought *material advantages* to a lot of people...but somewhere along the line...at different times in different places...it out-lived its usefulness...economic growth continued *but there was no corresponding increase in the satisfaction of people's needs...*

WHAT WE GOT WERE...

...bigger cities...bigger factories, offices, shops...a faster pace of work...more roads for more cars travelling longer distances to work...greater energy wastage...greater consumption of finite resources...

THIS 'STRUCTURAL TRANSFORMATION' ALSO BROUGHT US...

...more stress at work and alienation outside work...more social problems...more industrial and traffic accidents...more useless products...more artificial environments...more pollution...

WAIT, LISTEN: OUR ECONOMIC GROWTH ENABLES US TO TAKE CARE OF THOSE WHO CAN'T KEEP UP...

...BUT MORE AND MORE ARE FALLING BY THE WAYSIDE...

...SO TO TAKE CARE OF THEM WE HAVE TO STEP UP OUR ECONOMIC GROWTH...

...SO THAT WE CAN AFFORD TO LOOK AFTER MORE AND MORE WHO CAN'T STAND THE PACE...

...BECAUSE WE'VE HAD TO BOOST ECONOMIC GROWTH IN ORDER TO...

Ivan Illich believes social relations are bound to be degraded by high energy-use...

THERE'S AN ENERGY THRESHOLD ABOVE WHICH USAGE GROWS AT THE EXPENSE OF SOCIAL JUSTICE AND PROGRESS... BEYOND IT THE POLITICAL SYSTEM AND CULTURAL CONTEXT OF ANY SOCIETY MUST DECAY.

IF:

HIGH ENERGY CONSUMPTION = HIGH LIVING STANDARD = = PERSONAL HAPPINESS

THEN:

THIS WOMAN MUST BE FIVE TIMES AS HAPPY AS THIS WOMAN

BECAUSE THE AVERAGE CHINESE USES A FIFTH OF THE ENERGY USED BY THE AVERAGE AMERICAN.

OR:

THIS MAN'S STANDARD MUST BE FIVE TIMES HIGHER THAN THIS MAN'S

BECAUSE ROUGHLY THE SAME AMOUNT OF ENERGY GOES INTO MAKING LONG-LIFE FRIDGES AS ONES THAT BREAK DOWN QUICKLY. AND PORGY HAS TO KEEP BUYING NEW ONES...

109

⚡ ENERGY USE

A Swedish University study of the relationship between energy use and living standards reached some startling conclusions...

Two-thirds of the **total increase in energy use** since World War II had been devoured by the 'structural transformation' of society or simply **gone to waste!!**

Only a third of the increase could strictly be said to have improved the standard of living — better homes, hygiene and health...more leisure travel...construction of social and sports centres...and so forth.

EVEN IF WE COULD MEASURE OUR WELFARE BY THE AMOUNT OF ENERGY WE USE — MEASURING ENERGY CAN BE A MISLEADING BUSINESS...

AMORY LOVINS,
international energy expert:

66 How much *primary energy* we use — the fuel we take out of the ground — does not tell us how much energy is delivered at the *point of end use*...for that depends on how efficient our energy supply system is. *End-use energy* in turn says nothing about how much *function* we perform with the energy, for that depends on our end-use efficiency. And how much *function* we perform says nothing about *social welfare*, which depends on whether the thing we did was worth doing.**99**

AND MISUSE

The wastage referred to in the Swedish study is mostly what we lose in converting one form of energy (**eg coal**) into another (**eg electricity**) and distributing it. These conversion losses are often a result of the **wrong kind** of energy being used for the task in hand...

THE DISTINCTION BETWEEN PRIMARY ENERGY AND END-USE ENERGY IS VITALLY IMPORTANT...IF WE SEEK TO SUPPLY PRECISELY THE RIGHT AMOUNT OF ENERGY IN THE MOST DESIRABLE FORM FOR THE JOB TO BE DONE WE SOON FIND THAT GIGANTIC, INFLEXIBLE POWER UNITS ARE NOT WHAT WE NEED...

Electricity is not a fuel. It can't easily be stored in quantity. It loses two-thirds of the original energy in conversion. It's a **high-quality** form of power best suited for end-uses such as lighting, electronics, public transport and certain mechanical processes. In industrial countries these usually make up **less than 10% of total energy use...**

111

THE *ELECTRIC* FUTURE

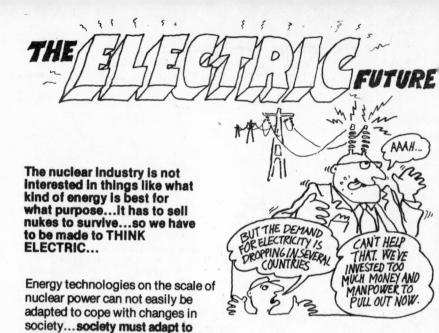

The nuclear industry is not interested in things like what kind of energy is best for what purpose...it has to sell nukes to survive...so we have to be made to THINK ELECTRIC...

Energy technologies on the scale of nuclear power can not easily be adapted to cope with changes in society...**society must adapt to them!** They commit us to a certain type of energy and a certain type of lifestyle...like it or not...

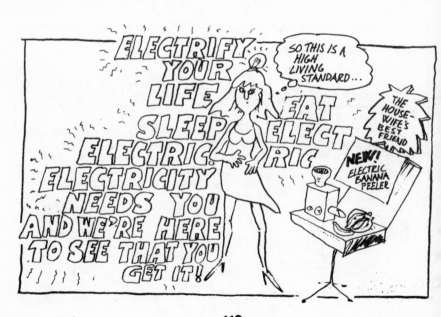

REACTORS AND JOBS...

US estimates show that big power stations produce fewer jobs per dollar than just about any other major investment! British estimates show that the electrical supply industry is the most capital-intensive of them all....

Amory Lovins:

66 It's the conservation, solar, environmental and related social programmes that yield the most energy, jobs and monetary returns per dollar invested...

The huge, capital-intensive energy facilities often proposed to relieve unemployment not only make it worse by draining from the economy the capital that could make more jobs if invested almost anywhere else, but also worsen inflation by tying up billions of dollars non-productively for a decade. 99

THAT'S HOW LONG IT TAKES TO BUILD A BIG POWER STATION THESE DAYS.

MORE ENERGY = MORE PRODUCTION = MORE EMPLOYMENT ... OR...

Up to now production and energy consumption have grown more or less in pace...but the connection is not automatic.

Today energy consumption in many places is static... and energy forecasts are looking less and less reliable. Estimates in the U.S. about future energy requirements vary as much as 500%...and in Britain:

MORE ENERGY = MORE PLANT = FEWER JOBS...

MORE AND MORE ENERGY HAS BEEN USED TO MAKE MORE EXPENSIVE AND AUTOMATED PLANT...

...RESULTING IN MORE AND MORE **STRUCTURAL** UNEMPLOYMENT... THOUSANDS OF JOBS LOST FOREVER.

WELL, YOU KNOW WHAT THEY SAY: GROW OR DIE!

WITH FEWER PEOPLE AND MORE PLANT AND ENERGY WE CAN PRODUCE MORE!

The capitalist economic system needs to sell more and more consumer goods...whether or not people can afford them...and they certainly can't afford them if they're on the dole...

WE MUST HAVE **SELECTIVE** GROWTH IN PRODUCING THINGS THAT MEET SOCIAL NEED INSTEAD OF GREED... AND THAT MEANS LABOUR-INTENSIVE PRODUCTION WHICH USES LESS ENERGY AND PROVIDES MORE SKILLED EMPLOYMENT.

WE MUST?

RABBLE-ROUSING NONSENSE!

TRADE UNIONS

DEPARTMENT OF ENERGY

THE LONG AND SHORT OF IT...

Abandoning the high-energy, hard technology approach is going to create *short-term* difficulties for certain groups...for instance workers making nuclear components. In some countries the bosses have jumped at this opportunity to drive a wedge between the trade unions and the anti-nuclear power movement...

But this divide-and-rule tactic is not always a success...

Rules of the Game
1. You make a product that's profitable but pollutes the environment.
2. You build another factory to make anti-pollution plant & equipment.
3. You sell them to your first factory.
4. You move to another country when the laws get too tough.

Scoring
Extra points for
* persuading the government to subsidise the clean-up…
* safeguarding your profits & your business reputation…
* saying you take the consequences of your actions…
* moving to the Third World and calling it foreign aid…

The Players
The Multinational Corporations
(who usually made the mess
in the first place)

STARVATION IS THE GREATEST ENVIRONMENTAL PROBLEM OF ALL.

Two-thirds of the global population is starving or close to starvation...

Much of the world's protein resources are used to raise livestock in the rich nations (10 kilos of cereal go into one kilo of beef)...

A third of the world's people — the rich nations — use four-fifths of the world's energy supply...

80% of food exports from the Third World go to the industrialised countries...

Much of the Third World's most fertile land is used to produce tobacco, tea, coffee, cotton and other colonial goods for the rich nations...

Raising the poor nations to today's US energy level would exhaust current oil reserves within a few years...

The Third World situation is a legacy of colonialism...the rich nations built their industrial development on stealing the poor nations' assets...killing and enslaving their peoples...smashing their agriculture and handicrafts...reducing them to 'supplier states' for the advancement of Western Civilisation...

UNTIL THEY CAME ALONG WE WERE DEVELOPING QUITE NICELY...

SOME OF OUR CIVILISATIONS WERE IN SOME WAYS MORE DEVELOPED THAN THE EUROPEAN ONES AT THE TIME...

IT WAS THE RIP-OFF CAPITALISTS WHO MADE US UNDER-DEVELOPED!

The legacy remains. Many Third World countries have their own flags and parliaments...but they're still 'foreign-owned'...in the 'neo-colonial' grip of the industrialised world...in other words, victims of IMPERIALISM...

TECHNICAL FIXES AND FOREIGN AID...

...ARE NOT MUCH USE...

...WITHOUT ECONOMIC AND SOCIAL CHANGE!

So it's a bit futile to imagine it'll all come right in the end if we just keep going as we are and give the poor a little more money and hardware...

TAKE CHINA AND INDIA FOR EXAMPLE...

Both were in roughly the same position in the 1940s...starvation...poverty...dependence...but what happened?

INDIA	CHINA
political liberation	political and economic liberation
lots of foreign aid	little foreign aid
no fundamental changes in the structure of production	many-sided development of industry and agriculture
advanced technology including nuclear power	emphasis on simple, small-scale technologies
green revolution — eclipse of many small farmers, mass use of chemical fertilisers and DDT for short-term benefits	red revolution — union of peasant, factory worker and technician, more careful use of resources

OUTCOME	
Continuing starvation, poverty and distress...still an under-developed country...	No starvation, considerable equality and no dependence...no longer an underdeveloped country...

A new economic order adjusting the balance of payments between governments is only half a solution.

Giving more money to the local rulers in Iran, Peru or Algeria — whether it's foreign aid, written-off debts or higher prices for raw materials — scarcely benefits the vast majority of citizens...

HIGH-ENERGY ECONOMIC GROWTH IN THE INDUSTRIALISED WEST IS BUILT ON IMPERIALISM — THE LOGICAL EXTENSION OF CAPITALISM — AND IT CAN'T BE FRIGHTENED AWAY BY SPEECHES IN THE U.N....

NO NUKES IS GOOD NUKES...

More and more developing countries are going for a system geared to their own resources and needs. It emphasises self-reliance, including 'self-help' technology. It draws on the experience of other countries but is based on the local conditions and traditions...

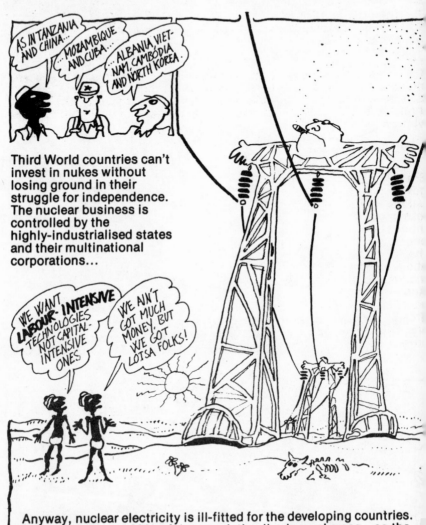

AS IN TANZANIA AND CHINA...

...MOZAMBIQUE AND CUBA...

...ALBANIA, VIET-NAM, CAMBODIA AND NORTH KOREA.

Third World countries can't invest in nukes without losing ground in their struggle for independence. The nuclear business is controlled by the highly-industrialised states and their multinational corporations...

WE WANT **LABOUR-INTENSIVE** TECHNOLOGIES, NOT CAPITAL-INTENSIVE ONES

WE AIN'T GOT MUCH MONEY, BUT WE GOT LOTSA FOLKS!

Anyway, nuclear electricity is ill-fitted for the developing countries. Between 80% and 90% of the population live in rural areas...so the overwhelming need is for **agricultural** not **industrial** development...

is. A glance at energy planning figures shows that the industrialised
ates are aiming to have *both nukes and oil!* If the growth-hungry West
 allowed to rush on down the hard technology road for another two or
ree decades there won't be much oil or gas left in the ground for those
ho need it most...

ome western
overnments are
lking vaguely about
velling off
dustrial growth and
ergy use. But
ey're *not* talking
bout abolition of the
arket economy.
nd the market
conomy demands
rowth...and more
nd more cheap
nergy...

QUESTION: WHAT HAPPENS WHEN THE GOVERNMENT OF A
CAPITALIST STATE TRIES TO INTRODUCE ZERO GROWTH?

123

SO IT SEEMS THAT THE HARD ENERGY ROAD
DOES **NOT** LEAD TO

BETTER LIVING STANDARDS...
MORE EMPLOYMENT...
GLOBAL EQUALITY...

ON THE CONTRARY...

FORTUNATELY
THERE IS AN ALTERNATIVE...

...FOR RICH AND
POOR NATIONS
ALIKE...

AND SO TO CHAPTER 7.

In which resources are husbanded, energy is renewable, decision-making is shared and a Soft Energy Future seems the only rational course...

THE ALTERNATIVE SOCIETY GOES UNDER MANY NAMES...

And takes many shapes. No single model fits everywhere. In fact, it's not so much a question of models as a strategy for change at various times with specific goals in mind...

Tactics vary. Some people think the profiteers can be reformed...that the ruling élites can be talked out of their privileges...that national parliaments can control multinational monopoly capitalists. Some think the environment crisis can be dealt with by juggling prices and tariffs...

But the idea is growing that only a socialist blueprint for change can reverse today's destructive trends...

THE ALTERNATIVE GOALS...

§ Switch-over from fossil fuels to renewable energy forms...
§ Conservation of natural and human resources...
§ Global solidarity...
§ Meaningful production and work for all...
§ Grass-roots government...

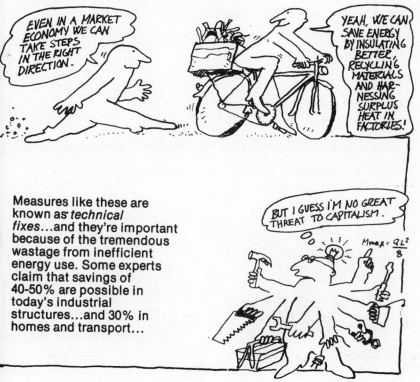

Measures like these are known as *technical fixes*...and they're important because of the tremendous wastage from inefficient energy use. Some experts claim that savings of 40-50% are possible in today's industrial structures...and 30% in homes and transport...

Changing over to the naturally-distributed energy sources is ecologically essential. But they can be developed in different ways...

THE MALIGNANT WAY
To prop up the centralised and authoritarian structures of our existing industrial society...

FOR EXAMPLE THERE ARE PLANS AFOOT FOR:

A vast solar power station in orbit beaming energy to Earth by microwave...

THE BENEVOLENT WAY
To promote the decentralisation of political and economic power, redistribution of wealth and the liberation of the individual...

A gigantic battery of 300,000 wind turbines on the American Great Plains...

SO IT'S NOT **JUST** A QUESTION OF HARD OR SOFT…HIGH OR LOW…BIG OR SMALL…

ANY ENERGY STRATEGY MUST ALSO BE JUDGED BY ITS POLITICAL AND ECONOMIC IMPLICATIONS…

WHAT ARE THE SOFT ENERGY SYSTEMS?

The strength of the 'soft energy path' is its flexibility...it's a mixture of the very many techniques directly suited to people's varying energy requirements...Lovins describes them like this:

> " They are diverse, so that as a national treasury runs on many small tax contributions, so national energy supply is an aggregate of very many individually modest contributions, each designed for maximum effectiveness in particular circumstances... "

They're especially suitable for Third World development because they directly satisfy the basic human needs...heating...cooking... lighting...pumping...

WHEN WILL THEY BE AVAILABLE?

The picture is changing fast... a few years ago we were told the soft energy systems were of marginal interest...now the Worldwatch Institute in Washington says sunlight can supply 40% of the world's needs by the turn of the century...and a British study says we can all be living on the renewable energy sources within about 75 years...

TODAY MANY GOVERNMENTS ARE EMBARKING ON ALTERNATIVE ENERGY PROGRAMMES...AND IT'S THE NUCLEAR INDUSTRY THAT'S HAVING TO CHASE FUNDS...

EXERCISE IN DOUBLE MORALITY

THE FOSSIL-FUEL BRIDGE...

To get to the stage where we can live on our energy **income** we'll have to keep drawing on our energy **savings** a while longer... half a century or more in the West...much less in the Third World...

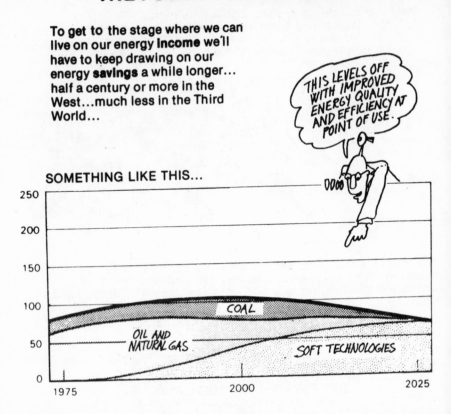

THIS LEVELS OFF WITH IMPROVED ENERGY QUALITY AND EFFICIENCY AT POINT OF USE.

SOMETHING LIKE THIS...

COAL

OIL AND NATURAL GAS

SOFT TECHNOLOGIES

Phasing out oil and natural gas means a *brief increase* in the use of coal, which is 25 times as plentiful. It has mainly been used to produce electricity and steel so far...but coal technology is undergoing a bit of a revolution these days and looks likely to replace oil in many respects...

MINING ACCIDENTS ARE GOING DOWN...

BUT WE STILL NEED A LOT MORE SAFETY MEASURES...

...AND REMOTE-HANDLING EQUIPMENT.

THE LONG-TERM AIM IS FULLY-AUTOMATED MINING!

THE FISSION-FUSION FANTASY...

The nuclear industry prefers another alternative — *a fission bridge*...which means the hard technology path...which means more and more energy...more wastage...more electricity creating a need for more nukes...breeders... super-breeders...and finally...

THE GREATEST NUKE OF THEM ALL

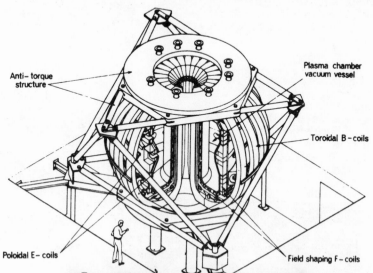

Anti-torque structure

Plasma chamber vacuum vessel

Toroidal B-coils

Poloidal E-coils

Field shaping F-coils

The FUSION REACTOR

It slams together atoms instead of splitting them. Its raw materials are lithium and deuterium. A fusion reactor would be colossal...starting at 3,000 MWe...and make fission reactors look cheap to build. But the industry says it will be more economical to run and less dangerous...

A CASE IN POINT

Motor-cars are just about the worst energy-wasters around…One remedy is to improve engines so that they use fuel more efficiently. Another is to make cars last 30, 40 or 50 years so you don't need to keep making new ones…

THESE ARE **FILTHY COMMIE IDEAS…** BUILT-IN OBSOLESCENCE KEEPS THE WHEELS TURNING!

NOT MINE!

Even better, do both and shift the emphasis from private motoring to public transport…less vehicles, less fuel, less raw materials, less energy…and less pollution!

BEST OF ALL: DECENTRALISE SOCIETY SO THAT PEOPLE LIVE WITHIN EASY DISTANCE OF THEIR FRIENDS AND WORKPLACES!

THAT'S A LONG-TERM GOAL…AND LIKE THE SWITCH TO COLLECTIVE TRANSPORT IT CAN SCARCELY BE ACHIEVED IN A MARKET ECONOMY…

lasting solution of the resources problem requires careful planning of energy use, work patterns and products...social equality requires humanised production under workers' and consumers' control...a system attuned to people's real needs...not shaped by the manipulated 'demands' of passive consumers...

BUT SUBSTITUTING A PLANNED ECONOMY FOR A MARKET ECONOMY IS NOT THE ULTIMATE GOAL...ONLY A PRECONDITION FOR REAL CHANGE...

In today's market economies science and technology is largely owned and controlled by the capitalist multinationals and serves their social, political and economic goals...

In tomorrow's planned economies research and development must be strongly focused on things like public transport, socially-useful products, soft energy techniques, recycling and safe workplaces...

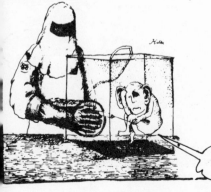

IF WE DON'T CONTROL TECHNOLOGY, IT WILL CONTROL US.

UNPLANNABLE ECONOMIES AND UNECONOMIC PLANNING...

Calls for some form of planning are now being heard from economic experts in such capitalist strongholds as Britain and the United States...without much success...

AND THAT GREY OLD KREMLIN CROWD ISN'T MUCH OF AN INSPIRATION!
The Soviet experiment went off the rails long ago...today it's more or less a parody of socialism...a centralised, authoritarian, bureaucratic and energy-hungry society riddled with nukes...

THE CHINESE EXAMPLE

Of existing societies, China has come closest to the soft energy model. Mao Tse-tung stressed the importance of local initiative and striking a balance between agriculture and industry, city and countryside, worker and student...

CONSERVING RESOURCES IN A PLANNED ECONOMY
The Disappearing Waste

The waste product from one factory is raw material for another...so you site the second industry next door from the start...and its surplus is used by a third...and so on...

For years, China has been a display window for alternative, ecologically-viable techniques...and has proved that people's individual and collective resourcefulness can work wonders in the proper setting....

WALKING ON TWO LEGS....

The Chinese policy of mixing big scale and small scale,
local and centralised, urban and rural, mental and manual,
can also be applied to industrialised societies...
approaching the problem from the other direction...
learning from our mistakes...

We can bury nukes, sophisticated weapons, supertankers,
supersonic airliners, chemical sprays and electric banana
peelers along with capitalism...while happily inheriting
useful things like computers, steam engines, cameras,
telephones, tape recorders, typewriters, printing presses,
bicycles, zeppelins, etc etc...

....
....
....
....

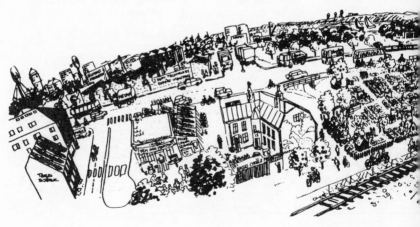

The practical circumstances are hard to foresee... But someone had a crack the other day in a British radical science journal, Undercurrents...

OFF THE TOP OF MY HEAD I CONSIDERED WHAT MIGHT BE ACHIEVED IN A PRAGMATIC, ECOLOGY-CONSCIOUS, ANTI-IMPERIALIST SOCIALIST SOCIETY FULLY CONFIDENT IN IT'S TECHNOLOGICAL CAPACITY BUT NOT OBSESSED WITH ADVANCED TECHNOLOGY. READY? TAKE A DEEP BREATH...

Large and small-scale sun, wind, water, geothermal etc energy usage. Collective digestion of all organic wastes for methane. Total energy systems and district heating. Heat pumps, hydrogen and methanol as basic fuels. Central electricity generation from many different sources with local grids. Fuel cells. Controlled exploitation of abundant resources. Advanced, autonomous houses outside the cities, mostly of large size. All the main industrial materials, although not in overwhelming abundance. Wide range of special steels and plastics. Semi-conductors and electronic bric-brac. Some computers, radio, telly etc. Much automation in tedious production jobs. Litho-presses. Proper R & D. Knowledge firmly based on the traditional sciences. Tractors. Combines etc. Machines for every [...]. Many more in [...] the f[...]duction. Health food a[...] mass food[...] tribution. Some meat. Trains. Public transport. Dirigibles. Few planes. Some long-distance travel but less commuting. Strong controls on environmentally harmful substances. Very strict emission standards. Less mining, with improved work conditions. Controlled distribution of raw materials. Careful conservation programme. Equipment designed for reliability, ease of repair, long life and recycling of components. Smaller range of consumer goods. Wider range of living patterns and possibility for greater variety within a person's life. Life-long education-and-work. Regional emphasis within a national and international economy. ETC...

EVERYONE RULES OK?

DRAFT PROPOSAL

1. Creation of committees or *councils* at local factories, housing estates and other places where people gather...

2. Council members directly elected and subject to immediate recall if they abuse responsibilities...

3. Local councils elect councils for larger units and regions...

4. Overall co-ordination...mmm...difficult...what about a directly-elected national planning body aided by council   working groups with specialised fields?

5. Factory councils decide the conditions of work — job safety, division of labour, rotation of work...

6. They cooperate with other council groups in determining the future course of production...

In living areas, neighbourhood councils could be responsible for housing management, local trading, day nurseries and recreation facilities...they could sound out local opinion on what consumer goods should be produced and distribute these in the district...

Civil rights are underpinned by widespread public involvement in mass communication, for example at community media centres...

COMMUNITY MEDIA CENTRE

CHOOSING THE SOFT ENERGY PATH AND RECOGNISING THE PHYSICAL AND BIOLOGICAL CONSTRAINTS ON HUMAN ACTIVITY DOESN'T MEAN FREEZING IN THE DARK...

It means choosing a *reasonable* material standard and a *reasonable* level of energy use instead of growing inequality and alienation...it means placing greater value on things that *count* instead of things that are merely *countable*...

We're all basically creative... It's only when we're not given responsibility or the tools to control our own lives that we become passive, violent, neurotic..

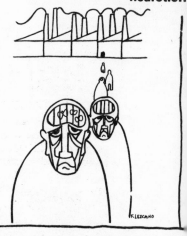

Active people who like to experiment and solve problems are not easy to oppress... they develop the self-confidence that is the lifeblood of real democracy...they don't need to be mobilised, led, ruled or watched over...quite literally, they're capable of taking matters into their own hands...

AND FINALLY: CHAPTER 8.

In which the environmentalists start the ball rolling, the workers tackle them, both revise their ideas and just about everyone joins the anti-nuclear struggle...

LIBERAL TRADE UNIONIST ECO-LOGIST RADICAL CONSER-VATION-IST CONCERNED CITIZEN CRITICAL SCIENTIST

HELL NO, WE WON'T GLOW...

Opponents of nuclear power come in many shapes and sizes...

They're more or less agreed on the dangers of nuclear technology and on the advantages of the soft energy path...but they adopt widely differing approaches to the problem...with varying degrees of success...

AS A SPECIES, ANTI-NUKERS FIRST CROPPED UP IN THE UNITED STATES IN THE EARLY 1960s. BY THE LATE 1970s THEY HAD SPREAD EN MASSE TO WESTERN EUROPE, JAPAN AND AUSTRALIA.

In the beginning, resistance mainly took the form of lobbying government or challenging the nuclear builders at licence hearings, public inquiries and in court...

Today, the emphasis is shifting more and more to direct, non-parliamentary methods and to building a mass movement against nuclear power...sometimes linking up across national boundaries...

PSST... OFF THE NUKE!

OFF THE NUKE!

RADIATION IGNORES BORDERS TOO...

144

Once, it was little more than a technical discussion between experts for the benefit of the politicians...

Now, the energy debate has become a matter for the general public, not just the decision-makers...and the citizenry isn't taking too kindly to being force-fed with nukes...

ALMELO, NETHERLANDS: 30,000 PROTEST...

MALVILLE, FRANCE: 50,000 PROTEST...

MELBOURNE, AUSTRALIA: 20,000 PROTEST...

NEW YORK UNITED STATES: 200,000 PROTEST...

MONTALTO, ITALY: 10,000 PROTEST...

BILBAO, SPAIN: 150,000 PROTEST...

BONN WEST GERMANY: 100,000 PROTEST...

LONDON, ENGLAND: 10,000 PROTEST...

Not to mention Sam Lovejoy...

66 On George Washington's Birthday, 1974 Samuel Holden Lovejoy toppled a 500-foot-tall weather tower in Montague, Massachussets. The tower had been erected by the local utility company as part of their project to construct one of the largest nuclear power plants ever planned. Leaving 349 feet of twisted wreckage behind, Lovejoy hitched a ride to the police station, where he turned himself in and submitted a four-page written statement decrying the dangers of nuclear power. Six months later, Lovejoy defended his act of civil disobedience in court as «self-defence». He was ultimately acquitted. **99**

The anti-nuclear power movement is growing militant with the realisation that you can't de-rail the Nuclear Express by gentle persuasion...

145

THE ECOLOGISTS POINTED THE WAY...

Environment groups launched the anti-nuke movement and have long been its vanguard...they were the first to spell out the dangers of a hard energy future...to organise resistance...and to formulate alternatives...

Adding an ecological dimension to the class struggle, radical environmentalists offered a fresh solution to the ravages of capitalism...but they also annoyed many dogged Marxists who kept insisting that technology was neutral...

BUT SOME ECOLOGISTS LOST THEIR WAY...

...AND MANY MISSED THE POINT...

SO IT WASN'T STRANGE THAT MUCH OF THE LABOUR MOVEMENT DISMISSED THE ENVIRONMENT MOVEMENT AS AN ASSEMBLY OF MIDDLE-CLASS CRAZIES OUT OF TOUCH WITH THE FACTS OF WORKING LIFE...

ECOLOGISTS HAD TO THINK AGAIN...

Their vision of an alternative society simply wasn't getting through...

...BUT SO DID THE WORKERS...

Under the constant threat of the capitalist axe, they'd been spending all their time battling for job security and better wages....

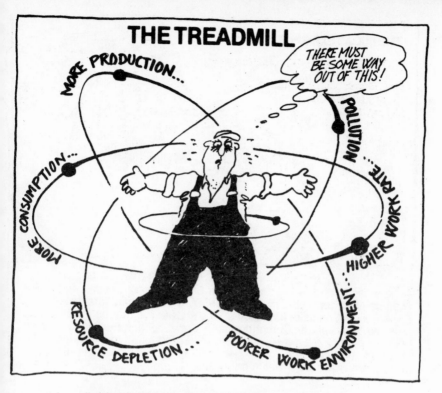

THE LUCAS EXAMPLE...

WHAT'S ALTERNATIVE PRODUCTION?

IT USUALLY MEANS MAKING THINGS THAT ARE SOCIALLY USEFUL AND NOT WASTEFUL!

LIKE KIDNEY MACHINES INSTEAD OF FIGHTER JETS

The workers and technicians at the British concern *Lucas Aerospace* campaigned for ALTERNATIVE PRODUCTION when threatened with closures and mass redundancies. The staff at the various Lucas plants formed a Combine Committee and drew up a full-scale Corporate Plan...proposing a move away from arms production to things like solar heating systems and wind generators, rubber-wheeled trains and insulation made from waste paper...

ALMOST ALL THE 150 PROPOSALS FOR ALTERNATIVE PRODUCTS CAME FROM THE WORKERS THEMSELVES!

A FEW CAME FROM OUTSIDE GROUPS LIKE ENVIRONMENTALISTS.

And the management's response?

The bosses refused to listen...which proved to everyone that reasoned argument makes no impression on vested interests... but news of Lucas workers' approach to the class struggle spread internationally...

TELL THEM NO BLOODY CHANGE DURING WORKING HOURS... ...AND IF THEY DON'T LIKE THAT, TELL THEM MILITARY AIRCRAFT ARE SOCIALLY USEFUL!

ALTERNATIVE PRODUCTION
took root as an idea...it was:

* discussed at workplaces from Lapland to Australia

* proposed for other industries like steel, pulp, packaging, car-making and shipbuilding

* raised in the British Parliament and the US Congress

* closely examined by trade journals and some of the mass media

A centre for the development of socially useful products in industry was set up in London in 1978. And Lucas Combine Committee speakers travelled into Europe at the invitation of unions there...

DEMANDS FOR ALTERNATIVE PRODUCTS AND MANUFACTURING METHODS BRING THE LABOUR MOVEMENT TO THE FOREFRONT OF THE ENVIRONMENT STRUGGLE...

Alternative production is not much good without alternative consumption! Socially useful products are useless if they don't sell!

The market economy is insensitive to social needs...but this must be *seen* to be the case! Groups outside the factory — environmentalists, consumers, unionists in the service sector — can help by bringing pressure on public institutions *to order the alternative goods to meet obvious needs...*

The Lucas Alternative Plan provided more than a meeting ground for the labour and environment movements...it also united manual workers and white-collar professionals...

Changing and improving industrial production is only half the battle...concrete demands must be made in many other areas of daily life...all pointing in the direction of a better social structure...

TOWARDS A TOTAL PERSPECTIVE...

In the Public Services...

In housing areas...

At the shops...

In public transport...

...AND A CO-ORDINATED DEMOCRATIC MOVEMENT!

quote those wise old ideological fathers of socialism Marx & Engels...

A REAL MOVEMENT...

... BEATS TWELVE PROGRAMMES!

Marx wasn't exactly an eco-freak. He did criticise capitalist agriculture and the dehumanising effect of separating people from nature...but he also tended to view capitalist technology as progressive. After him, Lenin, Trotsky, Stalin and other Marxists thought you could just take it over and paint it red

IF I'D KNOWN ABOUT FAST BREEDERS I MIGHT HAVE HAD SECOND THOUGHTS.

ANYWAY, MARXISM IS NOT JUST A THEORY... IT'S A WEAPON TO BE CONSTANTLY SHARPENED AND USED IN NEW WAYS.

But Marx & Engels were good at driving home one point — that all efforts to bring about fundamental social change must be mass-based if they're to succeed!

SO...

155

TALK TO FRIENDS...NEIGHBOURS...RELATIVES...FELLOW-
WORKERS...TAKE PART IN ACTIONS...JOIN GROUPS...GO TO
MEETINGS...GET ACTIVE IN YOUR UNION...OR YOUR LOCAL
COMMUNITY...OR BOTH...KEEP INFORMED...USE YOUR
IMAGINATION...

**BETTER ACTIVE TODAY
THAN RADIOACTIVE TOMORROW**

**IN THE END
IT'S UP TO YOU**

APPENDIX.

Alternative Energy Systems

The Sun...

Techniques harnessing the energy of the sun are being developed swiftly and widely. More than a century ago solar energy was used in Chile to desalinate water and in the US to power printing presses. Today it is in use in many fields and new applications are constantly being discovered.

FLAT-PLATE SOLAR COLLECTORS for domestic heating are the commonest hardware. They have long been in use in hot countries like Israel, South Africa and Australia. Now they are in great demand in the US, where sales are trebling annually.

Heating water in a West German swimming pool...and a Welsh cottage

Collectors are also finding a market in cloudier regions like Scandinavia and Britain as a source of hot water in summer and low-temperature heating in winter. The heat is usually stored in water or rockbeds. Solar collectors can be applied to both large and small premises, 'back-fitted' to existing buildings or more simply, fitted into the structure from the start. They are most economically used by groups of households sharing storage.

SOLAR FOCUSING COLLECTORS reflect and concentrate the sun's rays on one spot. They range from gigantic high-temperature furnaces capable of melting metals to do-it-yourself cookers for boiling water and baking bread, etc.

...ench solar furnace...Third World grill ...ade of bamboo, clay and foil...and a ...nk of solar [photo-voltaic] cells for the ...h

SOLAR CELLS turn sunshine into electricity. They are still extremely expensive but a major cost breakthrough is said to be imminent. In the US, government and private capital is being poured into solar cell technology, which lends self to monopoly. The same can be said of **SOLAR POWER TOWERS**, electric energy systems now being built in Sicily, France, Spain and the US. Ground mirrors provide the heat to turn a steam turbine in the tower.

The Wind...

An endless source of energy with no pollutive side-effects, the wind has been put to good use since 2000 BC when the Chinese and Persia ground corn with it. Current **WIND POWER** systems can heat, pump, compress air and generate electricity. Designs range from massive turbines on 50-metre towers geared to electricity grid supply...

Two of the World's Biggest...built by Danish school-kids and teachers at Tvind (left)... and by the US Space Administration in Ohio.

...down to the spinning oil drum variety for less ambitious ventures. The Greek island of Crete has over 100,000 small windmills pumping water for irrigation.

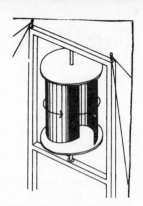

The Savonius Rotor...heavy but cheap and simple...and another vertical-axle wind generator, the high-speed Darrius which the Danes and Canadians think may produce 1.5 MWe...

Wind energy is highly versatile and an aid to decentralisation. It can be used in many ways at the point of use, reducing the need for long, expensive and wasteful powerlines. Battery storage is most common but there's talk of an 'inverter' that switches over to the grid when there's no breeze around...and also works the other way, turning your electric meter backwards!

British studies suggest that unit-for-unit windpower's total cost is one-third that of nuclear power...and the World Meteorological Office says 20 million MWe is blowing about...

But the wind does not always blow and the sun does not always shine...so a mixture of the two (above) is useful. Good insulation lessens the need for both by reducing a building's heat requirements.

161

The Water...

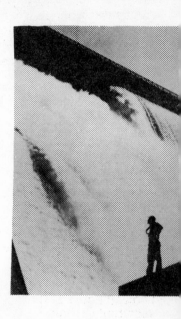

Harnessing the mechanical energy in falling water is the oldest technique in the book. The modern successor to yesterday's creaking water wheels is the streamlined turbine generating electricity, often in huge dam projects. These frequently damage the environment and disrupt local cultures. But it is no longer accepted that only large volumes of water can be economical and there is growing interest in medium-sized

HYDRO-ELECTRIC POWER PLANTS,

especially in Finland and the Soviet Union , while the plunging cost of small water turbines is making 'micro-stations' look attractive once more. Many countries have idle water wheels that can be revived to make electricity — there are 20,000 in England alone. China has built 60,000 in 15 years to supply 20% of her electricity needs. Most of the world's untapped water power is in the developing countries, especially in Africa, which has 20% of the global potential but generates a mere 2%.

Water power is the most reliable of the renewable energy sources. Except in extreme droughts it flows non-stop, strongest in winter when demand is greatest. However, most waterways these days are 'owned'...

TIDAL POWER SYSTEMS exploit the ebb and flow of the tides so are not dependent on climate. A small scale station is operating in France, a larger one is being built in Canada's Bay of Fundy and a giant tidal power barrage across Britain's River Severn is under consideration.

La Rance has been providing 240 MWe for 10 years...and two British wave-power designs with built-in turbines...

WAVE POWER techniques harnessing the energy of the ocean waves are mainly under development in Britain, Japan and Scandinavia. The Norwegians estimate that a 150-kilometre string of wave power units could supply 70,000 million kilowatt hours of electricity — the nation's current annual consumption.

OCEAN THERMAL POWER UNITS exploiting the difference in temperature between the sea depths and the surface are soaking up a lot of US government funds in anticipation of a mid-1980s market...

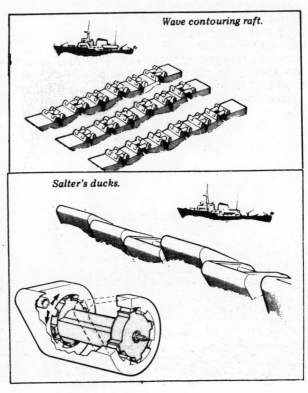

Wave contouring raft.

Salter's ducks.

The Earth...

Plants solved the energy crisis 3,000 years ago when they learned to create themselves out of earth, air and water. Now they're back in favour as 'biomass crops' or **BIO-FUELS.**

Apart from direct wood-burning, methods exist to tap the energy stored in plants, trees and organic waste by converting them into methane and other liquid and gaseous fuels for the transport sector. Special 'energy plantations' of fast-growing trees like alder and poplars (left) are being cultivated while ocean farming of giant seaweeds and other marine plants is also under way. Large-scale organic conversion presents some environmental problems like soil depletion and requires careful integration in forestry and agriculture.

Small-scale production of methane is a well-tested technology among Chinese and Indian farmers (right) whose air-tight digesters are mostly filled with cow-dung and human sewage. It is estimated that animal waste and crop residues on US farms could supply all the power needs of that country's enormous agricultural industry.

RECYCLING

...metals, glass, wood, paper, cloth, oil, batteries and even some plastics is another energy-saver...'waste' is a raw material in the wrong place at the wrong time!

GEO-THERMAL ENERGY,

drawing on the heat deep in the Earth's crust, is easiest to plumb in volcanic regions but will become available elsewhere with better drilling techniques. Another centralised, high-technology system for the energy corporations and their state backers.

HEAT PUMPS,

drawing on the earth's warmth or the air, are already in widespread use. Many are powered from the mains. New designs link heat pumps to wind and solar energy systems.

Recycling scrap metal...geo-thermally heating a Paris block of flats...and pump-heating a row of 12 Colorado apartments...

URANIUM RESERVES

HIGH-YIELD up to $80/kg cost-of-recovery
LOW-YIELD over $80/kg

| | (THOUSAND TONS) | |
	HY	LY
Algeria	28	—
Argentina	17.8	24
Australia	289	7
Brazil	18.2	—
Canada	167	15
France	37	14.8
Gabon	20	—
India	29.8	—
Niger	160	—
South Africa/Namibia	306	42
Sweden	—	300
United States	523	120
Portugal	6.8	—
Spain	6.8	—
WORLD TOTAL* (32 countries)	1,510	590

*No data available for socialist states

URANIUM PRODUCTION (1977)
(TONS)

Argentina	130
Australia	400
Canada	6,100
France	2,200
Niger	1,609
Portugal	85
South Africa	6,700
Spain	191
US	11,200

Total (14 non-socialist countries)
28,617 tons

ACTUAL AND ESTIMATED URANIUM REQUIREMENTS

Source: Int. Symposium on Uranium Supply and Demand, London July 1978

	1977	1980	1985	1990
Non socialist world. (rounded)	24,400	40,500	76,000	111,000

167

ORGANISATIONS AGAINST NUCLEAR POWER

Australia

Movement Against Uranium Mining (MAUM), 277 Brunswick St, Fitzroy, Victoria 3065.

Community Research Action Centre, Monash University Union, Clayton, Victoria 3168.

Friends of the Earth, 232 Castlereagh St, Sydney, New South Wales 2001.

Austria

Initiative Österreichischer Atomkraftwerksgegner (IOAG), Postfach 138, 1071 Vienna.

Belgium

Amis de la Terre, Rue E. Walschaert 19, 1060 Bruxelles.

Verenigde Aktiegroepen voor Kernstop (VAKS), Consciencestraat 46, 2000 Antwerpen.

Britain

Socialist Environment & Resources Association (SERA), 9 Poland St, London W1V 3DG.

Scottish Campaign to Resist the Atomic Menace (SCRAM), 2a Ainslyplace, Edinburgh 3.

Friends of the Earth, 9 Poland St, London W1V 3DG.

Nuclear Information Network (NIN), 29 Great James St, London WC1.

London Greenpeace, 6 Endsleigh St, London WC1.

Colonialism & Indigenous Minorities Research Action (CIMRA), 70 Durham Rd, London N7.

Canada

Canadian Coalition for Nuclear Responsibility (CCNR), 2010 Mackay St, Montreal, Quebec H3G 2J1.

Saskatoon Environmental Society, PO Box 1372, Saskatoon, Saskatchewan S7K 3N9.

Denmark

Organisationen til Oplysning om Atomkraft (OOA), Skindergade 26, DK-1159 Copenhagen K.

Finland

Alternative till Kärnkraft, PB 143, 00201 Helsinki 20.

France

Coordination Antinucléaire Région Parisienne, C/o Yves Behar, 3 rue Félix Ziem, 75018 Paris.

Comité contre la pollution dans la Hague, BP 156, 50104 Cherbourg Cédex.

Comité contre le Super-Phénix (Malville), C/o Marie Dubost, Iselet, Morestel.

Amis de la Terre, 117 Av. de Choisy, 75013 Paris.

Comité de Sauvegarde de Fessenheim et de la plaine du Rhin (CSFR), C/o J.J. Rettig, Ecole de Champenay, 67420 Saales.

West Germany

Bundersverband Bürgerinitiatieve Umweltschutz (BBU), Schiffkopfweg 31a, 75 Karlsruhe 21.

Burgerinitiatieve Umweltschutz Unterelbe (BUU), Schlüterstr. 4, 2000 Hamburg 13.

Anti-AKW-Laden, Lutterothstr. 33, 2000 Hamburg 19.

Bürgeraktion Küste (BAK), Bückerburgerstr. 50, Bremen 2800.

Gruppe Internationalismus in der Bremer Bürgerinitiative gegen Atomanlagen, Fedelhören 14, 28 Bremen 1.

Ireland

C/o John Carroll, Irish TGWU, Liberty Hall, Dublin 1.

Japan

Jishu Koza, Masafumi Takubo, B. Kaikan, Nishi Okubo 2-350, Shinjukuku, Tokyo.

Luxemburg

Bürgerinitiative Museldal, C/o Jemp Weydert, 51 rue demy Schlechter, Luxemburg.

Netherlands

Landelijk Energie Kommitee (LEK), 2e Weteringplantsoen 9, Amsterdam.

New Zealand

Friends of the Earth, Box 39-065, Auckland West.

Norway

Aksjon mot Atomkraft, PB 8395, Hammersborg, N. Oslo 1.

Spain

Comision de Defensa de una Costa Vasca no Nuclear, Avenida Basagoiti 28, Algorta.

Comité Antinuclear de Catalunya (CANC), Bruc 26 2on., Barcelona 10.

Sweden

Folkkampanjen mot Atomkraft, Tjärhovsgatan 44, 11629 Stockholm.

Miljöförbundet, Box 2129, 750 02 Uppsala.

Switzerland

National Koordination der Schweizer Anti-AKW-Organisationen, C/o André Froidevaux, Burgunderstr. 4, 4051 Basel.

Schweizerische Energiestiftung (SES), Auf der Mauer 6, 8001 Zurich.

United States

Abalone Alliance, 452 Higuerast, San Louis Obispo, California 93401.

Cactus Alliance, 106 Girard SE, Rm 121C, Albuquerque, New Mexico 87106.

Clamshell Alliance, PO Box 30, Montague Center, Massachusetts 01351, and 22 Congress St, Portsmouth, New Hampshire 03801.

Community Energy Action Network, PO Box 33686, San Diego, California 92103.

Mobilization for Survival, 1213 Race St, Philadelphia 19107.

Natural Resources Defence Council, 917 15th Street, Washington, D.C. 20005.

Friends of the Earth, 124 Spear Street, San Francisco, California 94105.

General

World Information Service on Energy (WISE), 2e Weteringplantsoen 9, Amsterdam, Netherlands.

International Mobilisation for Survival, C/o ICDP, 6 Endsleigh St, London WC1, Britain.

Remember him from page 54?

A news article by Bryan Silcock in the **Sunday Times,** 6 Jan. 1980, carried this headline: **Nuclear accident alters map of Russia.**

Here's the gist of it. New evidence has come to light about the vast scale of the nuclear accident in the Urals in 1957. This accident was reported in the West in 1976 by the exiled scientist Zhores Medvedev. He has since pieced together indirect evidence for the disaster from papers published in Soviet journals dealing with botany, biology, and genetics, described in his recent book, **Nuclear Disaster in the Urals.**

Medvedev's work stimulated the environmental sciences division at Oak Ridge National Laboratory, USA, to undertake an independent investigation. Oak Ridge compared the Soviet maps made of the area before 1958 and those made during the 1970s. These show:

1. that the names of 30 small communities with populations of under 2,000, as well as some larger towns, have been removed from the later maps.
2. In one 60-mile stretch of countryside, **all** the towns shown in the earlier maps have disappeared.
3. Significant changes appear in the lakes and river system of the area. For instance, Lake Kzyltash is now by-passed by a new canal system, and two big new reservoirs with a total area of 50 square kilometres have appeared. There is no obvious explanation for their construction in such a well-watered region.
4. Isolating Lake Kzyltash and the reservoirs from the Techa river system, according to Oak Ridge, "strongly indicates" that this is intended to prevent waterborne radioactive contamination, such as Strontium-90, from moving downstream in the Techa system.

What was the precise nature of the Urals disaster? Medvedev originally mentioned an explosion of radioactive wastes buried underground. But what could have caused an explosion powerful enough to spread radioactivity throughout such a wide area?

Oak Ridge scientists now suggest an explosion of the ammonium nitrate formed in the wastes. On reasonable assumptions, it might have had the force of 100 tons of TNT. This explosion might have set up a pattern of contamination which has literally changed the map of Russia!

170

Tips for further reading

Books

Berger, J. Nuclear Power: The Unviable Option. Dell, 1977.
Bookchin, M. Post-Scarcity Anarchism. Ramparts, 1971.
Breach, Ian. Windscale Fallout. Penguin Special, 1978.
Commoner, Barry. The Poverty of Power. Bantam, 1977.
Commoner, Barry. The Politics of Energy. Knopf, 1979.
Elliot, D. ed. The Politics of Nuclear Power. Pluto Press, 1978.
Flowers Report, The. Cmnd. 6618 HMSO London 1976.
Fuller, J. G. We almost Lost Detroit. Ballantine, 1976.
Harper, P. and Boyle, G. eds. Radical Technology. Pantheon, 1976.
Hayes, Denis. Rays of Hope: The Transition to a Post-Petroleum World. Norton, 1977.
Illich, Ivan. Energy and Equity. Marion Boyars, 1974.
Illich, Ivan. Tools for Conviviality. Calder and Boyars, 1973.
Junck, R. The Nuclear State. Calder, 1979.
Lovins, Amory. Soft Energy Paths. Penguin, 1977.
Patterson, W. The Fissile Society. Earth Resources, 1977.
Roberts and Medvedev. The Hazards of Nuclear Power. Spokesman, 1978.
Smith, D. ed. The Big Red Nuclear Diary. Pluto Press, 1978.
Union of Concerned Scientists. The Risks of Nuclear Power Reactors. UCS, Boston, 1977.

Papers, articles, comix...

Is Nuclear Power Necessary? A. Lovins. Friends of the Earth, 1979. 9 Poland Street, London W1.
'Nuclear Power — Who Needs it?' Science for the People Vol. 8 No. 3 May 1976.
The Leveller Energy Issue. No. 12. Feb 1978. 155a Drummond St, London NW1.
'Nuclear Power — No thanks!' FOE 1977. 9 Poland St, London W1.
Nuclear Times. FOE 1978.
An Alternative Energy Strategy for the UK. NCAT, Machynlleth, Powys, Wales.
All Atomic Comics. Last Gasp Educomics 1976. From 112 Fellows Rd, London NW3.

Periodicals

Undercurrents, 27 Clerkenwell Clo, London, EC1R OAT, Britain.
New Scientists, 128 Long Acre, London WC2E 9OH.
Science for People, 9 Poland St, London W1.
Science for the People, 9 Walden St, Jamaica Plain, MA 02130, USA.
Bulletin of the Atomic Scientists, 1020-24 East 58th St, Chicago, III 60637.
Environment, 438 N. Skinker Boulevard, St Louis, Mo. 63130.
Spark, CSRE 475 Riverside Drive, New York, NY 10027.
Radical Ecologist, PO Box 87, Carlton South, Victoria 30C3, Australia.
Chain Reaction, from FOE, 232 Castlereagh St, Sydney, NSW 2001, Australia.
Agenor, 13 Rue Hobbema, 1040 Brussels, Belgium.
La Gueule Ouverte, BP 26, 71800 La Clayette, France.
Survive et Vivre, 6 Rue Chappe, 75018, Paris, France.

BEGINNERS
FOR BEGINNERS

Series Editor Richard Appignanesi

Writers and Readers